TURING 高等院校计算机教材系列

计算机英语教程

吕云翔 杨雪 编著

人民邮电出版社

北 京

图书在版编目（CIP）数据

计算机英语教程 / 吕云翔，杨雪编著 . —北京：人民邮电出版社，2009.8
（高等院校计算机教材系列）
ISBN 978-7-115-19967-6

Ⅰ. 计… Ⅱ.①吕…②杨… Ⅲ. 电子计算机－英语－高等学校－教材 Ⅳ. H31

中国版本图书馆CIP数据核字（2009）第104198号

内 容 提 要

　　本书是面向计算机及相关专业的专业英语课程的教材，它全面介绍和讲解了深刻影响着我们生活的信息技术，内容既包含最新科研成果、业界前沿课题和发展趋势，又有计算机文化典故和名人轶事。本教材注重英语听说读写能力的全面发展和实际应用。各章节内容均分为阅读与翻译、写作、听说三大部分。通过听、说、读、写、译全方位训练，使读者掌握英语交流所应具备的基本技能及计算机相关知识。

　　本书适合国内各大院校信息技术、计算机、通信工程等专业教学之用，也可作为IT领域技术人员和管理层人员的自修参考用书。

高等院校计算机教材系列

计算机英语教程

◆ 编　　著　吕云翔　杨 雪

　　责任编辑　杨海玲

　　执行编辑　陆春凌

◆ 人民邮电出版社出版发行　　北京市崇文区夕照寺街14号

　　邮编　100061　　电子函件　315@ptpress.com.cn

　　网址　http://www.ptpress.com.cn

　　北京鑫正大印刷有限公司印刷

◆ 开本：800×1000　1/16

　　印张：17.5

　　字数：456千字　　　　　　2009年8月第1版

　　印数：1 - 3 000册　　　　　2009年8月北京第1次印刷

ISBN 978-7-115-19967-6/TP

定价：35.00元

读者服务热线：(010)51095186　印装质量热线：(010)67129223

反盗版热线：(010)67171154

前　言

作为全球IT行业的通用交流语言，英语是每一位IT从业人员必须掌握的语言，计算机专业英语知识的运用与实践是IT从业人员必须具备的基本职业技能。

本教材是参照全国计算机等级考试（计算机职业英语部分）的要求，按照最新《大学英语教学大纲》，为各类高校和职业学校开设的计算机英语课程而编写的，适用于计算机相关专业学生和广大从事计算机相关工作的在职人员。

本教材选材广泛，内容丰富。全书共分为12章，分别从计算机基础、计算机硬件、计算机软件、操作系统、计算机程序设计、数据库、计算机网络、因特网、电子商务、计算机安全、软件工程和计算机的未来等方面全面介绍和讲解深刻影响着我们生活的信息技术，内容既包含最新的科研成果、业界前沿课题和发展趋势，又有计算机文化典故和名人轶事。

本教材在对话场景的编排上以三位计算机专业大学本科生Mark、Henry和Sophie的学习生活为主要背景，他们交流的话题围绕各章主题展开，并在对话中丰富各章主题，将全书内容巧妙地联系在一起。

本教材信息容量大，知识性强，注重英语的听、说、读、写、译能力的全面培养和实际应用。各章内容均分为阅读与翻译、写作、听说三大部分。其中，阅读分为精读和泛读两部分，精读部分全面和丰富地论述本章主题，使读者深入了解和掌握相关专业知识；泛读部分介绍计算机领域的最新技术进展，供读者开阔视野；两部分均列出了计算机专业词汇。翻译部分结合阅读部分的文章，将其中涉及的复杂句型和特殊句型，或涉及计算机相关的重要知识点的句子摘录出来，一部分作为阅读提示，另一部分作为翻译练习，帮助读者巩固计算机和英语的专业知识。写作部分讲解IT常用文体写作方法，且在方法指导的基础上辅以实例。听说部分是与各章主题相关的专题讨论，将计算机的相关知识与实际的场景对话相结合，旨在综合训练读者的听说能力，并在对话中掌握计算机的相关知识。通过听、说、读、写和译全方位的训练，使读者掌握英语交流所应具备的基本技能，并学习计算机相关知识。

本教材采用场景式教学和体验式学习相结合的方式，教材中设计的听力、口语、阅读与翻译和写作练习融合了角色扮演、多人会话和小组讨论等行之有效的训练方法，能较好地满足课堂教学的需要。

　　另外，本教材配有配套的MP3听力材料，听力录音均聘请专业人员编录，可为学生提供非常有价值的口语模板。配套的MP3听力材料、练习参考答案以及全部正文的中文翻译可以在图灵网站www.turingbook.com本书网页免费注册下载。本教材建议教学时长为48学时或32学时（可根据情况进行适当的调整）。

　　本教材在编写的过程中，得到了美国专家Eric Langager的指导，王昕鹏和傅尔也参加了全书的审校工作，在此一并表示衷心的感谢。

　　本教材在编写和出版过程中得到人民邮电出版社的鼎力支持。

编　者

2009年7月于北京

目 录

Contents

The Fundamentals of Computers

Part 1   Reading and Translating

Section A   The Varieties of Computers

Before you begin, you will need to understand three key concepts.

1. Purpose of a Computer: Turn Data into Information

- Data: Data consists of the raw facts and figures that are processed into information—for example, the votes for different candidates being elected to student-government office.

- Information: Information is data that has been summarized or otherwise ***manipulated*** for use in decision making—for example, the total votes for each candidate, which are used to decide who won.

2. Difference between Hardware and Software

- Hardware: Hardware consists of all the machinery and equipment in a computer system. The hardware includes, among other devices, the keyboard, the screen, the printer, and the "box"—the computer or processing device itself. Hardware is useless without software.

- Software: Software, or programs, consists of all the electronic *instructions* that tell the computer how to perform a task. These instructions come from a software developer in a form (such as **CD**, or compact disk) that will be accepted by the computer. Examples are Microsoft Windows and Office XP.

3. The Basic Operations of a Computer (Regardless of Type and Size)

- Input operation: Input is whatever is put in ("input") to a computer system. Input can be nearly any kind of data—letters, numbers, symbols, shapes, colors, temperatures, sounds, pressure, light beams, or whatever raw material needs processing. When you type some words or numbers on a keyboard, those words are considered input data.

- Processing operation: Processing is the manipulation a computer does to transform data into information. When the computer adds 2 and 2 to get 4, that is the act of processing. The processing is done by the central processing unit—frequently called just the **CPU**—a device consisting of electronic circuitry that executes instructions to process data.

- *Storage* operation: Storage is of two types—temporary storage and permanent storage, or primary storage and secondary storage. *Primary storage, or **memory**, is the internal computer circuitry that temporarily holds data waiting to be processed. Secondary storage, simply called storage, refers to the devices and media that store data or information permanently.* A *floppy disk* or *hard disk* is an example of this kind of storage. (Storage also holds the software—the computer programs.)

- Output operation: Output is whatever is put out ("output") from the computer system—the results of processing, usually information. Examples of output are numbers or pictures displayed on a screen, words printed out on paper by a printer, or music *piped* over some loudspeakers.

- Communications operation: Though not all computers have communications ability, which offers an extension capability. In other words, it extends the power of the computer. With wired or wireless communications connections, data may be input from *afar*, processed in a remote area, stored in several different locations, and output in yet other places. However, you don't need communications ability to write letters, do calculations, or perform many other computer tasks.

All Computers, Great and Small: The Categories of Machines.

At one time, the idea of having your own computer was almost like having your own personal nuclear reactor. In those days, during the 1950s and 1960s, computers were enormous machines affordable only by large institutions. Now they come in a variety of shapes and sizes, which can be classified according to their processing power: **supercomputers**, **mainframe** computers, **workstations**, microcomputers, and microcontrollers. We also consider servers.

1. Supercomputers

Typically priced from $1 million to more than $350 million, supercomputers are high-capacity machines with thousands of processors that can perform more than several trillion calculations per second. These are the most expensive and fastest computers available. "Supers," as they are called, have been used for tasks requiring the processing of enormous volumes of data, such as doing the **census** count, forecasting weather, designing aircraft, modeling molecules, and breaking **encryption** codes. More recently they have been employed for business purposes—for instance, **sifting demographic** marketing information—and for creating film **animation**.

Supercomputers are still the most powerful computers, but a new generation may be coming that relies on **nanotechnology**(Nano means "one-billionth."), in which molecule-sized nanostructures are used to create tiny machines for holding data or performing tasks. *A biological nanocomputer, which would be made of* **DNA** *and could fit into a single human cell, would use* **DNA** *as its software and* **enzymes** *as its hardware; its molecular-sized circuits would be viewable only through a microscope*. Some believe that within 10 years computers with the size of a pencil eraser will be available that work 10 times faster than today's fastest supercomputer. Eventually nanotech could *show up* in every device and appliance in your life.

2. Mainframe Computers

The only type of computer available until the late 1960s, mainframes are water- or air-cooled computers that cost $5,000 ~ $5 million and vary in size from small, to medium, to large, depending on their use. Small mainframes ($5,000 ~ $200,000) are often called midsize computers; they used to be called minicomputers, although today the term is seldom used. Mainframes are used by large organizations—such as banks, airlines, insurance companies, and colleges—for processing millions of **transactions**. Often users access a mainframe by means of a **terminal**, which has a display screen and a keyboard and can input and output data but cannot by itself process data. Mainframes process billions of instructions per second.

3. Workstations

Introduced in the early 1980s, workstations are expensive, powerful personal computers usually used for complex scientific, mathematical, and engineering calculations and for **computer-aided design** and computer-aided manufacturing. Providing many capabilities comparable to those of midsize mainframes, workstations are used for such tasks as designing airplane **fuselages**, developing prescriptions drugs, and creating movie special effects. *Workstations have caught the eye of the public mainly for their graphics capabilities*, which are used to *breathe* *three-dimensional* life into movies such as The Lord of the Rings and Harry Potter. The capabilities of *low-end* workstations *overlap* those of *high-end desktop microcomputers*.

4. Microcomputers

Microcomputers, also called personal computers (**PC**s), which cost from $500 to over $5,000, can fit next to a desk or on a desktop or can be carried around. They either are stand-alone machines or are connected to a computer network, such as a local area network. A local area network (**LAN**) connects, usually by special cable, a group of desktop PCs and other devices, such as printers, in an office or a building.

Desktop PCs are microcomputers whose case or main *housing* sits on a desk, with keyboard in front and monitor (screen) often on top.

Tower PCs are microcomputers whose case sits as a "tower," often on the floor beside a desk, freeing up desk surface space. Some desktop computers, such as Apple's 2004 iMac, no longer have a boxy housing; most of the actual computer components are built into the back of the flat-panel display screen.

Notebooks computers, also called *laptop computers*, are light-weight portable computers with built-in monitor, keyboard, hard disk *drive*, battery, and **AC** *adapter* that can be plugged into an electrical outlets; they weigh anywhere from 1.8 to 9 pounds.

Personal digital assistants (**PDAs**), also called *handheld computers* or *palmtops*, combine personal organization tools—schedule planners, address books, to-do lists—with the ability in some cases to send emails and faxes. Some PDAs have touch-sensitive screens. Some also connect to desktop computers for sending or receiving information. (For now, we are using the word digital to mean "computer based.") The range of handheld wireless devices, such as multipurpose cellphones, has surged in recent years.

5. Microcontrollers

Microcontrollers, also called *embedded computers*, are the tiny, specialized *microprocessors* installed in "smart" appliances and automobiles. These microcontrollers enable microwave ovens, for example, to store data about how long to cook your potatoes and at what power setting. Recently microcontrollers have been used to develop a new universe of experimental electronic appliances—e-pliances. For example, they are behind the new single-function products such as digital cameras, MP3 players, and organizers, which have been developed into *hybrid* forms such as *gadgets* that store photos and videos as well as music. They also help run tiny Web servers embedded in clothing, jewelry, and household appliances such as refrigerators. In addition, microcontrollers are used in blood-pressure monitors, air bag sensors, gas and chemical sensors for water and air, and vibration sensors.

6. Servers

A *server*, or network server, is a central computer that holds collections of data (databases) and programs for connecting or supplying services to PCs, workstations, and other devices, which are called *clients*. These clients are linked by a wired or wireless network. The entire network is called a client/server network. In small organizations, servers can store files, provide printing stations and transmit emails. In large organizations, servers may also house enormous libraries of financial, sales, and product information.

 Words

manipulate[mə'nipjuleit] *v.* 操作，处理	**transaction**[træn'zækʃ ən] *n.*事务，交易
instruction[in'strʌkʃ ən] *n.*指令	**terminal**['tə:minl] *n.*终端
storage['stɔridʒ] *n.*存储，存储器	**fuselage**['fju:zila:ʒ] *n.*飞机机身
memory['meməri] *n.*内存	**breathe**[bri:ð] *v.*将…注入
pipe[paip] *v.*传送	**overlap**['əuvə'læp] *v.*重叠，覆盖
afar[ə'fa:] *adv.*遥远地	**housing**['hauziŋ] *n.*壳
supercomputer[,sju:pəkəm'pju:tə] *n.*超级计算机	**drive**[draiv] *n.*驱动器
mainframe['meinfreim] *n.*主（计算）机	**adapter**[ə'dæptə] *n.*适配器
workstation['wə:ksteiʃ ən] *n.*工作站	**palmtop**[pa:mtɔp] *n.*掌上型电脑
census['sensəs] *n.*人口普查	**microcontroller**[maikrəukən'trəulə] *n.*微控制器
encryption[in'kripʃ ən] *n.*加密术，密码术	**microprocessor**[maikrəu'prəusesə] *n.*微处理器
sift[sift] *v.*筛选	**hybrid**['haibrid] *adj.*混合的
demographic[di:mə'græfik] *adj.*人口统计学的	**gadget**['gædʒit] *n.*小配件，机械装置，小工具
animation[,æni'meiʃ ən] *n.*动画片，卡通	**server**['sə:və] *n.*服务器
nanotechnology['nænəutek'nɔlədʒi] *n.*纳米技术	**client**['klaiənt] *n.*客户机
enzyme['enzaim] *n.*酶	

 Phrases

floppy disk	软盘	desktop microcomputer	台式微型计算机
hard disk	硬盘	notebook computer	笔记本电脑
show up	出现	laptop computer	膝上型电脑
three-dimensional	立体的，三维的	handheld computer	掌上电脑
low-end	低端的	embedded computer	嵌入式计算机
high-end	高端的		

Abbreviations

CD	Compact Disk	光盘
CPU	Central Processing Unit	中央处理器
DNA	DeoxyriboNucleic Acid	脱氧核糖核酸
CAD	Computer-Aided Design	计算机辅助设计
PC	Personal Computer	个人计算机
LAN	Local Area Network	局域网
AC	Alternating Current	交流电
PDA	Personal Digital Assistant	个人数字助理

Complex Sentences

1. **Original**: Tower PCs are microcomputers whose case sits as a "tower," often on the floor beside a desk, freeing up desk surface space.
   Translation: 塔式个人计算机是一种机箱像"塔"一样放置的微型计算机，通常机箱放在桌子旁边的地上，腾出了桌面空间。

2. **Original**: Workstations have caught the eye of the public mainly for their graphics capabilities.
   Translation: 主要由于其所具有的图形能力，工作站已经引起了公众的关注。

Exercises

I. Read the following statements carefully, and decide whether they are true (T) or false (F) according to the text.

____1. Computers can transform the raw data into information with meaning.

____2. Supercomputers calculate very fast with a high-capacity processor.

____3. Terminal is a specific type of mainframe.

____4. In a client/server network, the server sits in the geographic center of the entire network.

____5. CPU is one of the most vital components in processing operation in a computer.

II. Choose the best answer to each of the following questions.

1. By which criterion the computers are classified according to this article?

   (A) Figure   (B) Capability   (C) Price   (D) Size

2. "After entering your new password and pressing the 'submit' button, a message 'Password Changes Successfully!' shows up on the screen of your computer." In the above scenario, which kind of operation is NOT included?

   (A) Input   (B) Processing   (C) Storage   (D) Communication

3. Which kind of computer may be the most appropriate used in an automatic gate in a

household garage according to this article?

(A) Workstation　　(B) Microcomputer　(C) Microcontroller　(D) Server

III. Translating.

1. Primary storage, or memory, is the internal computer circuitry that temporarily holds data waiting to be processed. Secondary storage, simply called storage, refers to the devices and media that store data or information permanently.

2. A biological nanocomputer, which would be made of DNA and could fit into a single human cell, would use DNA as its software and enzymes as its hardware; its molecular-sized circuits would be viewable only through a microscope.

⟫⟫ Section B　The History of Computers

While computers are now an important part of the lives of human beings, there was a time when computers did not exist. Knowing the history of computers and how much progress has been made can help you understand just how complicated and innovative the creation of computers really is.

Unlike most devices, the computer is one of the few inventions that do not have one specific inventor. Throughout the development of the computer, many people have added their creations to the list required to make a computer work. Some of the inventions extend the types of computers, while others help computers to be further developed.

⊖ The Beginning

Perhaps the most significant date in the history of computers is the year 1936. It was in this year that the first "computer" was developed. It was created by Konrad Zuse and *dubbed* the Z1 Computer. *This computer stands as the first as it was the first system to be fully programmable.* There were devices prior to this, but none had the computing power that sets it apart from other electronics.

It wasn't until 1942 that any business saw profit and opportunity in computers. This first company was called ABC computers, owned and operated by John Atanasoff and Clifford Berry. Two years later, the Harvard Mark I computer was developed, *furthering* the science of computing.

Over the course of the next few years, inventors all over the world began to search more into the study of computers, and how to improve upon them. Those next ten years say the introduction of the *transistor*, which would become a vital part of the inner workings of the computer, the **ENIAC** I computer, as well as many other types of systems. The ENIAC 1 is perhaps one of the most interesting, as it required 20,000 *vacuum tubes* to operate. It was a massive machine, and started the revolution to build smaller and faster computers.

The age of computers was forever altered by the introduction of International Business Machines, or **IBM**, into the computing industry in 1953. This company, over the course of computer history, has been a major player in the development of new systems and servers for public and private use. *This introduction brought about the first real signs of competition within computing history, which helped to* **spur** *faster and better development of computers.* Their first contribution was the IBM 701 **EDPM** Computer.

A Programming Language Evolves

A year later, the first successful high level programming language, FORTRAN, was created. This was a programming language not written in *"assembly"* or *binary*, which are considered as very low level languages. FORTRAN was written so that more people could begin to program computers easily.

The year 1955, the Bank of America, *coupled with* Stanford Research Institute and General Electric, saw the creation of the first computers for use in banks. The **MICR**, or Magnetic Ink Character Recognition, coupled with the actual computer, the **ERMA**, was a *breakthrough* for the banking industry. It wasn't until 1959 that the pair of systems was put into use in actual banks.

In 1958, one of the most important breakthroughs in computer history occurred, the creation of the *integrated circuit*. This device, also known as the *chip*, is one of the base requirements for modern computer systems. *On every motherboard and card within a computer system, are many chips that contain information on what the boards and cards do.* Without these chips, the systems as we know them today cannot function.

Gaming, Mice & the Internet

For many computer users now, games are a vital part of the computing experience. 1962 saw the creation of the first computer game, which was created by Steve Russel and **MIT**, which was dubbed Spacewar.

The mouse, one of the most basic components of modern computers, was created in 1964 by Douglass Engelbart. It obtained its name from the "tail" leading out of the device.

One of the most important aspects of computers today was invented in 1969. **ARPA** net was the original Internet, which provided the foundation for the Internet that we know today. This development would result in the evolution of knowledge and business across the entire planet.

It wasn't until 1970 that Intel entered the *scene* with the first dynamic **RAM** chip, which resulted in an explosion of computer science innovation.

On the heels of the RAM chip was the first microprocessor, which was also designed by Intel. These two components, in addition to the chip developed in 1958, would **number** among the core components of modern computers.

A year later, the floppy disk was created, gaining its name from the flexibility of the storage unit. This was the first step in allowing most people to transfer bits of data between unconnected computers.

The first *networking card* was created in 1973, allowing data transfer between connected computers. This is similar to the Internet, but allows for the computers to connect without use of the Internet.

Household PC's Emerging

The next three years were very important for computers. This is the time when companies began to develop systems for the average consumers. The Scelbi, Mark-8 Altair, IBM 5100, Apple I and II, TRS-80, and the Commodore Pet computers were the **forerunners** in this area. *While expensive, these machines started the trend for computers within common households.*

One of the most major breakthroughs in computer software occurred in 1978 with the release of the VisiCalc Spreadsheet program. All development costs were paid for within a two-week period of time, which makes this one of the most successful programs in computer history.

1979 was perhaps one of the most important years for the home computer users. This is the year that WordStar, the first word processing program, was **released** to the public for sale. This drastically altered the usefulness of computers for the everyday users.

The IBM home computer quickly helped revolutionize the consumer market in 1981, as it was affordable for home owners and standard consumers. 1981 also saw the mega-giant Microsoft enter the scene with the **MS-DOS** *operating system*. This operating system utterly changed computing forever, as it was easy enough for everyone to learn.

The Competition Begins: Apple vs. Microsoft

Computers saw yet another vital change during the year of 1983. The Apple Lisa computer was the first with a graphical user interface, or a **GUI**. Most modern programs contain a GUI, which allows them to be easy to use and pleasing for the eyes. This marked the beginning of the out-dating of most text-based only programs.

Beyond this point in computer history, many changes and alterations have occurred, from the Apple-Microsoft wars, to the developing of microcomputers and a variety of computer breakthroughs that have become an accepted part of our daily lives. Without the initial first steps of computer history, none of those would have been possible.

Words

dub[dʌb] v.授予称号	**breakthrough**[ˈbreikˈθruː] n.突破，重大成就
further[ˈfəːðə] v.促进，推动	**chip**[tʃip] n.芯片
transistor[trænˈzistə] n.晶体管	**scene**[siːn] n.舞台
spur[spəː] v.刺激，激励，鞭策	**number**[ˈnʌmbə] v.列入，把…算作
assembly[əˈsembli] n.汇编，集合	**forerunner**[ˈfɔː,rʌnə] n.先驱
binary[ˈbainəri] n.二进制	**release**[riˈliːs] v.发布

 Phrases

vacuum tube	真空管，电子管	**on the heels of**	紧跟着
be coupled with	和…联合，结合	**networking card**	网卡
integrated circuits	集成电路	**operating system**	操作系统

 Abbreviations

ENIAC	Electronic Numerical Integrator And Calculator[Computer]	电子数字积分计算机
IBM	International Business Machines	（美国）国际商用机器公司
EDPM	Electronic Data Processing Machine	电子数据处理机
MICR	Magnetic Ink Character Recognition	磁墨水字符识别
ERMA	Electronic Recording Method of Accounting	电子账目记录方法
MIT	Massachusettes Institute of Technology	（美国）麻省理工学院
ARPA	Advanced Research Projects Agency	（美国国防部）高级研究计划署
RAM	Random-Access Memory	随机访问存储器
MS-DOS	Microsoft Disk Operating System	微软磁盘操作系统
GUI	Graphical User Interface	图形用户界面

Complex Sentences

1. **Original**: This computer stands as the first as it was the first system to be fully programmable.
   Translation: 因为这台计算机是第一个完全可编程的系统，所以它名列第一。
2. **Original**: On every motherboard and card within a computer system, are many chips that contain information on what the boards and cards do.
   Translation: 在计算机系统中的每一个主板和卡上，有许多包含信息的芯片，主板和卡执行这些芯片上的信息。

Exercises

I. Read the following statements carefully, and decide whether they are true (T) or false (F) according to the text.

____1. Before 1936, all devices which had the computing power had to rely on other electronics.

____2. IBM was the first company who initiated the computer business.

____3. ENIAC 1 was made with the technology of the vacuum tube.

____4. Prior to the high level programming language, 'assembly' or binary is difficult to write and understand for the majority.

____5. Apple produced the first computer with a GUI.

II. Choose the best answer to each of the following questions.

1. Which of the following descriptions is WRONG according to this article?

(A) It was the creation of the integrated circuit that made the volume of the computers smaller.

(B) ARPA net provided the foundation for the Internet that we use today.

(C) Intel produced the first chip in the world, which exploded the innovation of computer science.

(D) "Mouse" obtained its name from the feature of its figure.

2. Among the following products, which appeared the earliest during the computer software development history?

(A) MS-DOS   (B) WordStar   (C) Spacewar   (D) VisiCalc Spreadsheet

3. In the following devices, which is NOT included in the list of the core components of modern computers?

(A) Chip   (B) Transistor   (C) RAM   (D) Microprocessor

III. Translating.

1. This introduction brought about the first real signs of competition within computing history, which helped to spur faster and better development of computers.

2. While expensive, these machines started the trend for computers within common households.

Part 2 **Simulated Writing: Memo**

⊕Introduction

The business memorandum is essentially an "internal" business letter. The "memo" as it is usually called, is the key internal communications tool in most businesses and institutions. In fact, the business memo is arguably the most important communications instrument in an organization.

Memoranda are used to: announce, inform, advise, quantify, delegate, direct, discipline, instruct, request and transmit. In short, if something needs to be communicated and/or recorded formally in an organization, it is done with a memo. Memos are often routed, posted, and forwarded, which means they can reach a lot of people quickly. Effects of careless mistakes compound quickly, since they tend to generate even more memos asking for clarification. Memos also get filed, which means they can come back to haunt you later.

In fact, "memo" comes from the Latin memorandum, "a thing which must be remembered." While a memo generally requests or delivers a quick response to a specific question, it may also be a compact version of a short report, progress report, or lab report. Although section titles may appear awkward in a very short memo, they allow your readers to scan efficiently and respond quickly.

Typical Components

1. Heading Segment

Begin the memo with a heading segment. Make sure you address the reader by her or his correct name and job title. Courtesy titles are not necessary but make sure you spell everyone's names properly and don't use informal nicknames.

Use a job title after your name, and write your initials by your name. This confirms that you take responsibility for the contents of the memo.

Be specific and concise in your subject line. For example, "computers" could mean anything from a new purchase of computers to a mandatory software class for employees. Instead use something like, "Turning Computers off at Night." This also makes filing and retrieving the memo easy.

2. Opening Segment

Begin writing your memo by stating the problem—that is, what led to the need for the memo. Perhaps a shipment has not arrived, a scheduled meeting has been canceled, or a new employee is starting tomorrow.

After stating the problem, indicate the purpose clearly: Are you announcing a meeting, welcoming a new employee, or asking for input on adopting a new policy about lunch hour length?

3. Discussion Segment

As you write the business memo, in the discussion segment, give details about the problem. Don't ramble on incessantly, but do give enough information for decision makers to resolve the problem. Describe the task or assignment with details that support your opening paragraph (problem).

4. Closing Segment

After the reader has absorbed all of your information, close with a courteous ending that states what action you want your readers to take. Should they email their reports rather than hand in hard copies? Attend a meeting? Chip in for someone's birthday cake? A simple statement like, "Thank you for rinsing the coffeepot after pouring the last cup" is polite and clearly states what action to take.

Traditionally memos aren't signed. However, it is becoming more common for memos to close the way letters do, with a typed signature under a handwritten signature. Follow your company's example for this.

Except for memos that are essentially informal reports or instructional documents, write a business memo no more than one page long. In a memo, less is more.

Format

To further clarify your meaning, keep these business memo formatting ideas in mind:

- Headings help the readers skim for sections of the document. In general, you should use an arrangement like the following:

MEMORANDUM

TO: Direct addressee
FROM: Signatory (Remember, this is not necessarily the author of the document)
DATE: Month, day, year(Spell out)
SUBJECT: Key words to describe the content of the memo

- Numbered and bulleted lists make information easy to scan. Be careful to make lists parallel in grammatical form.
- Font sizes, underlining, bolding, and italicizing make headings and important information stand out.
- As in all technical and business communications, long paragraphs of dense texts make reading more difficult. Therefore, keep your paragraphs short and to the point.

Sample

Memorandum

To: All Students in Software School
From: Grace Li, Teaching Assistant of Computer Concepts
Date: July 20, 2008
Subject: New Procedure for Handing in Homework

There is a new procedure for handing in homework. I believe that you will find it an improvement on the old one. It will be a convenient procedure. The new procedure is as follows:

1. Pack your homework files as a .zip or .rar file.
2. Name your .zip or .rar file with the following format:
   Your student ID_Your name_Homework 1.

3. Login ftp:// 202.112.115.219 with your student ID.
4. Open the folder named Computer Concepts.
5. Open the sub-folder named First Homework.
6. Copy your named file and paste there.

Attention
1. Make sure that the format is requested.
2. Hand in within 3 days since the homework is assigned, otherwise, the folder will be closed.

Part 3   Listening and Speaking

⫸ Dialogue: Buying a New Desktop Computer

(After class, Sophie & Henry are standing by the door, waiting for Mark.)

Henry	Excuse me, Sophie. May I ask you some questions about computers?
Sophie	Sure. What can I do for you?
Henry	I want to buy a new desktop computer, but I'm not sure which kind is better, *brand name computer* or *self-made computer*[1]. What's your idea about it?
Sophie	Let me see. In my view, as complete sets of computers made by some professional computer manufacturers, brand name computers have passed proper **compatible** tests, have higher quality assurance, and have better after-sales services. [2]Probably, a brand name computer is better.
Henry	But it seems that self-made computer is more flexible and often much cheaper.
Sophie	That's right. But the biggest problem is that it will take much time to choose all of the components with appropriate types. [3]On the contrary, if choosing a brand name computer, you would not worry about problems such as components fitting and hardware maintenance within a warranty period. However, you have to tolerate its lack of customization for meeting your special requirements.[4]

Henry	If I assemble the computer by myself, how can I find a series of appropriate components in a short time?
Sophie	Err... I've not much experience on this issue. Maybe you can find some information on the Internet or ask our other friends.

(When they are talking, Mark comes toward them.)

Sophie & Henry	Hi, Mark.
Mark	Hi, Henry and Sophie.
Sophie	You are just on time. Henry can't decide which kind of computer to buy, brand name computer or self-made computer. What's you opinion on it?
Mark	Self-made computer, also known as **DIY** computer, composed by components from the same or different hardware vendors. Its greatest advantage is that users can choose all of those components according to their different requirements.[5] Do you have any special requirements for your new computer?
Henry	Nothing special. Only is the speed I think.
Sophie	Then, which component is the most significant for fast speed in the computer?
Mark	CPU, main board, RAM, graphics card, hard disk, and so on. In fact, all of those can impact the speed of the computer.
Henry	Oh, it sounds complex.
Mark	Actually, even though you buy a brand name computer, you can still change its components for your potential requirements in the future if necessary.
Sophie	*Above all*, I'm afraid that you have to take a *tradeoff* on conven- ience and customization.[6] However, it's up to you, Henry.
Henry	Ok, I've got it. Thanks very much for your valuable suggestions.
Mark & Sophie	You are welcome.

Exercises

Work in pairs, and make up a similar conversation by replacing the scenarios with other material in the below.

● Buying a New Laptop Computer

[1] The graphics cards in laptop computer can be sorted into two main classes: *integrated graphics cards* and *discrete graphics cards*.

[2] The advantages of discrete graphics cards:
More powerful, having higher performance, and can offer better display effect.

[3] The disadvantages of integrated graphics cards:
Poorer performance when coming to 3D graphics, because they use system memory instead of *dedicated* graphics memory.

[4] The disadvantages of discrete graphics cards:
More expensive, generating more heat and demanding better heat dispersing.

[5] The advantages of integrated graphics cards:
Cheaper in expense and lighter in weight for a *portable* cause as a laptop.

[6] The performance of the graphics card will directly influence both the frame rate and image quality of 3D programs and games. There are huge differences between the low-end and high-end cards in this respect. So a graphics card should be purchased *in accordance with* your needs.
If 2D is *as far as* you want to go, then you should look for a low-cost solution, perhaps even go for integrated graphics cards, which are perfectly fine if all you ever want from your PC is to use productivity applications and *surf* the Web.
If you have a demand of high 3D graphics performance, then you should choose a high-end solution, discrete graphics cards may be a better choice than integrated graphics cards.

 Words

compatible[kəm'pætəbl] *adj.*兼容的	**portable**['pɔ:təbl] *adj.*便携式的，易携带或移动的
dedicated['dedikeitid] *adj.*专用的	**surf**[sə:f] *v.*冲浪

 Phrases

brand name computer	品牌计算机	**integrated graphics card**	集成显卡
self-made computer	组装计算机	**discrete graphics card**	独立显卡
above all	最重要的是，首先	**in accordance with**	依照
tradeoff	折衷，权衡	**as far as**	直到，远到

 Abbreviations

DIY	Do It by Yourself	自己动手

⋙ Listening Comprehension: Roadrunner

Listen to the passage and the following 3 questions based on it. After you hear a question, there will be a break of 10 seconds. During the break, you will decide which one is the best answer among the four choices marked (A), (B), (C) and (D).

Questions

1. Who is the builder of the former No.1 on the TOP 500 list before Roadrunner's birth?
2. What does the word "hybrid" mean according to this article?
3. How long will it take for Roadrunner to complete a calculation that would have cost the 1998 machine 20 years to finish?

Choices

1. (A) Sony   (B) IBM   (C) AMD   (D) Red Hat
2. (A) Roadrunner combines its components made by different vendors.
   (B) Roadrunner can be used in a variety of applications.
   (C) Roadrunner is built by engineers and scientists from multi-countries.
   (D) Roadrunner's chips are made from different chemical materials.
3. (A) One hour   (B) Half a day   (C) One day   (D) Seven days

 Words

> **terabyte**['terəbait] *n.* 1000GB，万亿字节
>
> **bill**[bil] *v.* 宣布，公告
>
> **petaflop**[petəfləp] *n.* 每秒千万亿次浮点运算 (FLOP: Floating point Operations Per second)
>
> **trillion**['triljən] *num.* 万亿

⋙ Dictation: John von Neumann

This passage will be played THREE times. Listen carefully, and fill in the blanks with the words you have heard.

John von Neumann (1903—1957) was unquestionably one of the most *brilliant* scientists of the _____1_____. He was born in Budapest, Hungary, in 1903. In 1930 he joined the Princeton Institute for Advanced Study. He became a US _____2_____ in 1937, and during the Second World War *distinguished* himself with his work in weapons development. In 1955 he was named a Commissioner of the Atomic Energy Commission, a position he held up to his death from cancer in 1957.

Von Neumann made major contributions to *quantum mechanics* and _____3_____ physics and is perhaps best known for his work in the _____4_____ of computers during his all-too-short

_____5_____ in computer science since 1943. In the now famous **EDVAC** report of 1945, von Neumann clearly stated the idea of a _____6_____ program that *resides* in the computer's _____7_____ along with the data it was to _____8_____ on.

Instead of the ENIAC-the first _____9_____ computer *unveiled* in 1946-having program instructions *rewired* in for each new _____10_____ (typically requiring a half-day at least to prepare the machine for operation), stored-program computer kept its specific instructions in its _____11_____ , storing the information in the same _____12_____ as it would store any other information. To this end, the computer would necessarily contain five basic components: a _____13_____ , memory, a _____14_____ (CPU), and input and output components for interacting with human users. The _____15_____ would *delve* into memory, finding an instruction or _____16_____ , and deal with what it found _____17_____ . Stored-program computer was an _____18_____ over and was far more flexible than its *predecessor*. Moreover, its_____19_____ has become the prototype of most of its *successors*, including the_____20_____ of modern computers which exist to this day. Von Neumann's name has also become *synonymous* with modern computer architecture.

 Words

brilliant['briljənt] *adj.*超群的，杰出的	**rewire**[riːˈwaiə] *v.*重接电线
distinguish[disˈtiŋgwiʃ] *v.*使杰出，使著名	**delve**[delv] *v.*挖掘
quantum[ˈkwɔntəm] *n.*量子，量子论	**predecessor**[ˈpriːdisesə] *n.*前任，（被取代的）原有事物
mechanics[miˈkæniks] *n.*力学	**successor**[səkˈsesə] *n.*后继者，后续的事物
reside[riˈzaid] *v.*居住	**synonymous**[siˈnɔniməs] *adj.*同义的
unveil[ʌnˈveil] *v.*公布	

 Abbreviations

EDVAC	Electronic Discrete Variable Automatic Computer	离散变量自动电子计算机

Computer Hardware

Part 1   Reading and Translating
   Section A   Computer Motherboard
   Section B   Multi-core Processors
Part 2   Simulated Writing: Notices
Part 3   Listening and Speaking
   Dialogue: Referring to Websites or Online Forum for
   Microsoft Developer
   Listening Comprehension: Intel
   Dictation: Father of the Mouse - Doug Engelbart

Part 1   Reading and Translating

⋙ Section A   Computer Motherboard

A ***motherboard*** is the central or primary printed circuit board (**PCB**) making up a complex electronic system, such as a modern computer. It is also known as a mainboard, baseboard, system board, ***planar*** board, or, on Apple computers, a logic board, and is sometimes abbreviated casually as mobo.

Most motherboards produced today are designed for so-called IBM-compatible computers, which held over 96% of the global personal computer market in 2005. Motherboards for IBM-compatible computers are specifically covered in the PC motherboard articles.

A motherboard, like a ***backplane***, provides the electrical connections by which the other components of the system communicate, but unlike a backplane also contains the central processing unit and other subsystems such as real time clock, and some peripheral interfaces.

A typical desktop computer is built with the microprocessor, main memory, and other essential components on the motherboard. Other components such as external storage, controllers for video display and sound, and peripheral devices are typically attached to the motherboard via *edge* connectors and cables, although in modern computers it is increasingly common to integrate these "peripherals" into the motherboard.

Components and Functions

The motherboard of a typical desktop consists of a large printed circuit board. It holds electronic components and interconnects, as well as physical connectors (*sockets*, *slots*, and *headers*) into which other computer components may be inserted or attached.

Most motherboards include, at a minimum:
- sockets (or slots) in which one or more microprocessors (CPUs) are installed
- slots into which the system's main memory is installed (typically in the form of **DIMM** modules containing **DRAM** chips)
- a *chipset* which forms an interface among the CPU's **front-side bus**, main memory, and peripheral buses
- *non-volatile* memory chips (usually Flash **ROM** in modern motherboards) containing the system's *firmware* or **BIOS**
- a clock generator which produces the system clock signal to *synchronize* the various components
- slots for *expansion cards* (these interface to the system via the buses supported by the chipset)
- power connectors and circuits, which receive electrical power from the computer power supply and distribute it to the CPU, chipset, main memory, and expansion cards.

Additionally, nearly all motherboards include logic boards and connectors to support commonly-used input devices, such as PS/2 connectors for a mouse and keyboard. Early personal computers such as the Apple II or IBM PC included only this minimal peripheral support on the motherboard. Occasionally video interface hardware was also integrated into the motherboard; for example, on the Apple II, and rarely on IBM-compatible computers such as the IBM PC. Additional peripherals such as disk controllers and *serial ports* were provided as expansion cards. With the steadily declining costs and size of integrated circuits, it is now possible to include support for many peripherals on the motherboard. By combining many functions on one PCB, the physical size and total cost of the system may be reduced; highly-integrated motherboards are thus especially popular in small *form factor* and *budget* computers.

Given the high thermal design power of high-speed computer CPUs and components, modern motherboards nearly always include heat *sinks* and mounting points for fans to *dissipate* excess heat.

Temperature and Reliability

Motherboards are generally air-cooled with heat sinks often mounted on larger chips, such as the northbridge, in modern motherboards. Passive cooling, or a single fan mounted on the power supply, was sufficient for many desktop computer CPUs until the late 1990s; since then, most have required CPU fans mounted on their heat sinks, due to rising clock speeds and power consumption. Most motherboards have connectors for additional case fans as well. Newer motherboards have integrated temperature sensors to detect motherboard and CPU temperatures, and controllable fan connectors which the BIOS or operating system can use to regulate fan speed.

Some small form factor computers and home theater PCs designed for quiet and energy-efficient operation *boast* fan-less designs. This typically requires the use of a low-power CPU, as well as careful *layout* of the motherboard and other components to allow for heat sink *placement*.

*A 2003 study found that some **spurious** computer crashes and general reliability issues, ranging from screen image **distortions** to **I/O** read/write errors, can be attributed not to software or peripheral hardware but to aging **capacitors** on PC motherboards*. Ultimately this was shown to be the result of a faulty ***electrolyte formulation***.

Motherboards use electrolytic capacitors to filter the **DC** power distributed around the board. *These capacitors age at a temperature-dependent rate, as their water based electrolytes slowly evaporate*. This can lead to loss of ***capacitance*** and subsequent motherboard malfunctions due to voltage instabilities. While most capacitors are rated for 2000 hours of operation at 105℃, their expected design life roughly doubles for every 10℃ below this. At 45℃ a lifetime of 15 years can be expected. This appears reasonable for a computer motherboard, however many manufacturers have delivered substandard capacitors, which significantly reduce this life expectancy. Inadequate case cooling and elevated temperatures easily ***exacerbate*** this problem. It is possible, but tedious and time-consuming, to find and replace failed capacitors on PC motherboards; it is less expensive to buy a new motherboard than to pay for such a repair.

History

Prior to the advent of the microprocessor, a computer was usually built in a card-cage case or mainframe with components connected by a backplane consisting of a set of slots themselves connected with wires; in very old designs the wires were discrete connections between card connector *pins*, but printed-circuit boards soon became the standard practice. The central processing unit, memory and peripherals were housed on individual printed circuit boards which plugged into the backplane.

During the late 1980s and 1990s, it became economical to move an increasing number of peripheral functions onto the motherboard. In the late 1980s, motherboards began to include single ICs (called Super I/O chips) capable of supporting a set of low-speed peripherals: keyboard, mouse, floppy disk drive, serial ports, and *parallel ports*. *As of* the late 1990s, many personal computer motherboards support a full range of audio, video, storage, and networking functions without the need for any expansion cards at all; higher-end systems for 3D gaming and computer graphics typically retain only the graphics card as a separate component.

The early pioneers of motherboard manufacturing were Micronics, Mylex, AMI, DTK, Hauppauge, Orchid Technology, Elitegroup, DFI, and a number of Taiwan-based manufacturers.

Popular personal computers such as the Apple II and IBM PC had published *schematic* diagrams and other documentation which permitted rapid *reverse-engineering* and *third-party* replacement motherboards. Usually intended for building new computers compatible with the *exemplars*, many motherboards offered additional performance or other features and were used to upgrade the manufacturer's original equipment.

Bootstrapping Using the BIOS

Motherboards contain some non-volatile memory to initialize the system and load an operating system from some external peripheral device. Microcomputers such as the Apple II and IBM PC used read-only memory chips, mounted in sockets on the motherboard. *At power-up the central processor would load its program counter with the address of the boot ROM and start executing ROM instructions displaying system information on the screen and running memory checks*, which would in turn start loading memory from an external or peripheral device (*disk drive*) if one isn't available, so that the computer can perform tasks from other memory stores or display an error message depending on the model and design of the computer and version of the BIOS.

Most modern motherboard designs use a BIOS, stored in a **EEPROM** chip *soldered* to the motherboard, to bootstrap the motherboard. (Socketed BIOS chips are widely used also.) By *booting* the motherboard, the memory, circuitry, and peripherals are tested and *configured*. This process is known as a Power On Self Test or **POST**. Errors during POST result in POST error codes, ranging from simple audible *beeps* from the speaker to complex diagnostic messages displayed on the video monitor.

The BIOS often requires configuration settings to be stored on the motherboard. Since configuration settings must be easily edited, these settings are often stored in non-volatile RAM (**NVRAM**) rather than in some sort of read-only memory (ROM). When a user makes configuration changes or alters the date and time of the computer, this small NVRAM circuit stores the data. Typically, a small, long-lasting battery (e.g. a **_lithium_** coin cell CR2032) is used to keep the NVRAM "refreshed" for many years. Therefore, a failing battery on a motherboard will produce the symptoms of a computer that cannot determine the correct date and time, nor remember what hardware configuration the user has selected. The BIOS itself is unaffected by the status of the battery.

When IBM first introduced the PC in the 1980s, imitations were quite common. (The physical parts which made up the motherboard were trivial to acquire.) However, the imitations were never successful until the IBM ROM BIOS was legally copied. To understand why copying the BIOS was an important step, consider that the BIOS contained vital instructions which interacted with peripherals. Without these software instructions in the BIOS, a PC would not function properly. (In most modern computer operating systems, the BIOS is **_bypassed_** for most hardware functions, but in the 1980s, the BIOS served many vital low-level functions.)

So when Compaq Computer Corp. spent US$1 million to **_clone_** the IBM BIOS using reverse engineering, they became an **_elite_** computer manufacturer of IBM PC Clones. Phoenix Technology soon matched their feat and began reselling BIOSes to other clone makers. *It has been noted that Microsoft was more than happy to license the operating system (DOS), and IBM was more than happy to sue companies that violated the copyright of their BIOS.* But by documenting and publicizing the reverse engineering of the BIOS, Compaq and Phoenix were legally competing with IBM using their own copyrighted BIOS.

Once the bootstrapping of the computer's peripherals is complete, the BIOS will normally pass control to another set of instructions stored on a bootable device.

Devices which are normally used to boot a computer:
- floppy drive
- network controller
- **CD-ROM** drive
- **DVD-ROM** drive
- **SCSI** hard drive
- **IDE**, **EIDE**, or **SATA** hard drive
- External **USB** memory storage device

Any of the above devices can be stored with machine code instructions to load an operating system or a program.

Form Factors

Motherboards are produced in a variety of sizes and shapes ("form factors"), some of which are specific to individual computer manufacturers. However, the motherboards used in IBM-compatible *commodity* computers have been standardized to fit various case sizes. As of 2007, most desktop computer motherboards use one of these standard form factors even those found in Macintosh and Sun computers which have not traditionally been built from commodity components.

Laptop computers generally use highly integrated, miniaturized, and customized motherboards. This is one of the reasons that laptop computers are difficult to upgrade and expensive to repair. Often the failure of one laptop component requires the replacement of the entire motherboard, which is usually more expensive than a desktop motherboard due to the large number of integrated components.

 ## Words

motherboard[ˈmʌðəbɔːd] *n.*主板，母板	**capacitor**[kəˈpæsitə] *n.*电容器
planar[ˈpleinə] *adj.*平面的	**electrolyte** [iˈlektrəulait] *n.*电解质，电解液
backplane[bækplein] *n.*底板	**formulation**[ˌfɔːmjuˈleiʃən] *n.*由…制定的配方
edge[edʒ] *n.*边缘	**capacitance**[kəˈpæsitəns] *n.*电容，电容量
socket[ˈsɔkit] *n.*孔，插座	**exacerbate**[eksˈæsəbeit] *v.*恶化，加剧
slot[slɔt] *n.*插槽	**pin**[pin] *n.*针，管脚，引线
header[ˈhedə] *n.*制造（钉头）的工具	**schematic**[skiˈmætik] *adj.*示意性的
chipset[tʃipset] *n.*芯片集	**exemplar**[igˈzemplə] *n.*样本，示例
non-volatile[ˈnɔnˈvɔlətail] *adj.*非易失性的	**bootstrap**[ˈbuːtstræp] *v.*引导
firmware[ˈfəːmˌwɛə] *n*固件（软件硬件相结合）	**solder**[ˈsɔldə] *v.*焊接
synchronize[ˈsiŋkrənaiz] *v.*同步	**boot**[buːt] *v.*引导
budget[ˈbʌdʒit] *adj.*合算的，廉价的	**configure**[kənˈfigə] *v.*配置
sink[siŋk] *n.* 吸附物，槽	**beep**[biːp] *n.*嘟嘟声
dissipate[ˈdisipeit] *v.*驱散	**lithium**[ˈliθiəm] *n.*锂
boast[bəust] *v.*以有…而自豪	**bypass**[ˈbaipɑːs] *v.*忽视，绕过，回避
layout[ˈleiˌaut] *n.*布置，安排	**clone**[kləun] *v.*复制，克隆
placement[ˈpleismənt] *n.*放置，布局	**elite**[eiˈliːt] *n.*精锐，杰出人物
spurious[ˈspjuəriəs] *adj.*伪造的，假造的	**commodity**[kəˈmɔditi] *n.*日用品，商品
distortion[disˈtɔːʃən] *n.*变形，失真	

 ## Phrases

expansion card	扩展卡	**reverse-engineering**	逆向工程
serial ports	串行端口	**third-party**	第三方
form factor	外形	**power-up**	开机
parallel ports	并行端口	**disk drive**	磁盘驱动器
as of	到…为止		

Abbreviations

PCB	Printed Circuit Board	印制电路板
DIMM	Dual In-line Memory Module	双列直插式内存模块
DRAM	Dynamic Random Access Memory	动态随机访问存储器
FSB	Front Side Bus	前端总线
ROM	Read-Only Memory	只读存储器
BIOS	Basic Input Output System	基本输入输出系统
I/O	Input/Output	输入/输出
DC	Direct Current	直流电
EEPROM	Electrically Erasable Programmable Read-Only Memory	电可擦除可编程只读存储器
POST	Power On Self Test	通电自检测试
NVRAM	Non-Volatile RAM	非易失性随机访问存储器
CD-ROM	Compact Disc Read-Only Memory	光盘驱动器
DVD-ROM	Digital Video Disc Read-Only Memory	数字化视频光盘驱动器
SCSI	Small Computer Systems Interface	小型计算机系统接口
IDE	Integrated Drive Electronics	电子集成驱动器
EIDE	Enhanced IDE	增强型IDE接口
SATA	Serial Advanced Technology Attachment	串行高技术附件
USB	Universal Serial Bus	通用串行总线

Complex Sentences

1. **Original:** These capacitors age at a temperature-dependent rate, as their water based electrolytes slowly evaporate.

   Translation: 因为电容器基于水的电解液会慢慢地蒸发，所以它们老化的速度依赖温度。

2. **Original:** It has been noted that Microsoft was more than happy to license the operating system (DOS), and IBM was more than happy to sue companies that violated the copyright of their BIOS.

   Translation: 广为人知的是，微软公司很乐于授权DOS操作系统，而IBM公司非常热衷于对那些侵犯其BIOS版权的公司提出诉讼。

Exercises

I. **Read the following statements carefully, and decide whether they are true (T) or false (F) according to the text.**

___1. Early personal computer motherboards need expansion cards to support a set of high-speed peripherals.

___2. A motherboard provides physical sockets and slots as some peripheral interfaces.

___3. BIOS is stored in non-volatile memory chips.

___4. By combining many functions on one PCB, highly-integrated motherboards are more complex and more expensive.

___5. Motherboards designed for IBM computers held over 96% of the global personal computer market in 2005.

II. Choose the best answer to each of the following questions.

1. Which of the following is NOT attached to the motherboard via edge connectors and cables in a typical desktop computer?

(A) External storage   (B) Controllers for video display

(C) Microprocessors   (D) Peripheral devices

2. Which statement is RIGHT about the BIOS according to this article?

(A) In most modern computer operating systems, the BIOS serves most hardware functions.

(B) A BIOS is designed to bootstrap the motherboard for memory, circuitry, and peripherals testing and configuration.

(C) Once the bootstrapping of the computer's peripherals is complete, the BIOS will normally in turn control another set of instructions stored on motherboard.

(D) Compaq Computer Corp. and Phoenix Technology became elite computer manufacturers of IBM PC Clones by cloning the IBM BIOS.

3. Which statement is WRONG about the motherboard according to this article?

(A) Since printed-circuit boards became the standard practice, the central processing unit, memory and peripherals were housed on one printed circuit board which plugged into the backplane.

(B) Although many personal computer motherboards today can support a set of high-speed peripherals without the need for any expansion cards at all, higher-end systems retain the graphics card as a separate component for some specific functions.

(C) Now, desktop computer motherboards in Macintosh and Sun computers also use one of the standard form factors.

(D) Now, the BIOS or operating system can use controllable fan connectors integrated in motherboards to regulate fan speed.

III. Translating.

1. A 2003 study found that some spurious computer crashes and general reliability issues, ranging from screen image distortions to I/O read/write errors, can be attributed not to software or peripheral hardware but to aging capacitors on PC motherboards.

2. At power-up the central processor would load its program counter with the address of the boot ROM and start executing ROM instructions displaying system information on the screen and running memory checks.

⫸ Section B Multi-core Processors

A multi-core CPU (or chip-level multiprocessor, **CMP**) combines two or more independent cores into a single package composed of a single integrated circuit (IC), called a *die*, or more dies packaged together. A dual-core processor contains two cores, and a *quad*-core processor contains four cores. A multi-core microprocessor implements multiprocessing in a single physical package.

A processor with all cores on a single die is called a *monolithic* processor. Cores in a multi-core device may share a single coherent *cache* at the highest on-device cache level (e.g. L2 for the Intel Core 2) or may have separate caches (e.g. current AMD dual-core processors). The processors also share the same interconnect to the rest of the system. Each "core" independently implements optimizations such as *superscalar* execution, *pipelining*, and *multithreading*. A system with n cores is effective when it is presented with n or more *threads* concurrently. The most commercially significant (or at least the most "obvious") multi-core processors are those used in personal computers (primarily from Intel and AMD) and game consoles (e.g., the eight-core Cell processor in the PS3 and the three-core Xenon processor in the Xbox 360). In this context, "multi" typically means a relatively small number of cores. However, the technology is widely used in other technology areas, especially those of embedded processors, such as network processors and **digital signal processors**, and in **GPUs**.

The amount of performance gained by the use of a multi-core processor depends on the problem being solved and the *algorithms* used, as well as their implementation in software (Amdahl's law). For so-called "embarrassingly parallel" problems, a dual-core processor with two cores at 2GHz may perform very nearly as fast as a single core of 4GHz. Other problems though may not yield so much *speedup*. *This all assumes however that the software has been designed to take advantage of available parallelism. If it hasn't, there will not be any speedup at all.* However, the processor will *multitask* better since it can run two programs at once, one on each core.

⊝ Terminology

There is some *discrepancy* in the *semantics* by which the terms multi-core and dual-core are defined. Most commonly they are used to refer to some sort of central processing unit (CPU), but are sometimes also applied to digital signal processors (DSP) and system-on-a-chip (**SoC**). Additionally, some use these terms to refer only to multi-core microprocessors that are manufactured on the same integrated circuit die. These people generally refer to separate microprocessor dies in the same package by another name, such as multi-chip module, double core, or even twin core. This article uses both the terms "multi-core" and "dual-core" to reference microelectronic CPUs manufactured on the same integrated circuit, unless otherwise noted.

In contrast to multi-core systems, the term multi-CPU refers to multiple physically separate processing units (which often contain special circuitry to facilitate communication between each other).

Advantages

*The **proximity** of multiple CPU cores on the same die allows the cache coherency circuitry to operate at a much higher clock rate than is possible if the signals have to travel off-chip.* Combining equivalent CPUs on a single die significantly improves the performance of *cache snoop* (alternative: *bus snoop*) operations. Put simply, this means that signals between different CPUs travel shorter distances, and therefore those signals degrade less. These higher quality signals allow more data to be sent in a given time period since individual signals can be shorter and do not need to be repeated as often.

Assuming that the die can fit into the package, physically, the multi-core CPU designs require much less Printed Circuit Board (PCB) space than multi-chip **SMP** designs. Also, a dual-core processor uses slightly less power than two coupled single-core processors, principally because of the decreased power required to drive signals external to the chip and because the smaller silicon process **geometry** allows the cores to operate at lower voltages; such reduction reduces latency. Furthermore, the cores share some circuitry, like the L2 cache and the interface to the front side bus (FSB). In terms of competing technologies for the available silicon die area, multi-core design can make use of proven CPU core library designs and produce a product with lower risk of design error than devising a new wider core design. Also, adding more cache suffers from diminishing returns.

Disadvantages

In addition to operating system (**OS**) support, adjustments to existing software are required to maximize utilization of the computing resources provided by multi-core processors. Also, the ability of multi-core processors to increase application performance depends on the use of multiple threads within applications. The situation is improving.

Integration of a multi-core chip drives production **yields** down and they are more difficult to manage thermally than lower-density single-chip designs. Intel has partially **countered** this first problem by creating its quad-core designs by combining two dual-core on a single die with a unified cache, hence any two working dual-core dies can be used, as opposed to producing four cores on a single die and requiring all four to work to produce a quad-core. From an architectural point of view, ultimately, single CPU designs may make better use of the silicon surface area than multiprocessing cores, so a development **commitment** to this architecture may carry the risk of obsolescence. Finally, raw processing power is not the only constraint on system performance. Two processing cores sharing the same system bus and memory **bandwidth** limit the real-world performance advantage. If a single core is close to memory bandwidth limits, going to dual-core might only give 30% to 70% improvement. However, if memory bandwidth is not a problem, a 90% improvement can be expected. *It would be possible for an application that used two CPUs to end up running faster on one dual-core if communication between the CPUs was the limiting factor, which would count as more than 100% improvement.*

Hardware Trend

The general trend in processor development has been from multi-core to many-core: from dual-, quad-, eight-core chips to ones with tens or even hundreds of cores. In addition, multi-core chips mixed with simultaneous multithreading, memory-on-chip, and special-purpose "***heterogeneous***" cores promise further performance and efficiency gains, especially in processing multimedia, recognition and networking applications. There is also a trend of improving energy efficiency by focusing on performance-per-watt with advanced *fine-grain* or ultra fine-grain power management and dynamic voltage and frequency scaling (**DVFS**), which is of particular interest for mobile computing (i.e. laptop computers and *portable media players*).

Software Impact

Software benefits from multi-core architectures where code can be executed in parallel. Under most common operating systems this requires code to execute in separate threads or ***processes***. Each application running on a system runs in its own process so multiple applications will benefit from multi-core architectures. Each application may also have multiple threads but in most cases, it must be specifically written to utilize multiple threads. Operating system software also tends to run many threads as a part of its normal operation. Running virtual machines will benefit from adoption of multiple core architectures since each virtual machine runs independently of others and can be executed in parallel.

Most application software is not written to use multiple concurrent threads intensively because of the challenge of doing so. A frequent pattern in multithreaded application design is that a single thread does the intensive work while other threads do much less. For example, a ***virus*** scan application may create a new thread for the scan process while the GUI thread is waiting for commands from the user (e.g. cancel the scan). In such cases, multi-core architecture is of little benefit for the application itself due to the single thread doing all heavy lifting and the inability to balance the work evenly across multiple cores. Programming truly multithreaded code often requires complex co-ordination of threads and can easily introduce subtle and difficult-to-find ***bugs*** due to the ***interleaving*** of processing on data shared among threads (thread-safety). Consequently, such code is much more difficult to ***debug*** than single-threaded code when it ***breaks***. There has been a perceived lack of motivation for writing consumer-level threaded applications because of the relative rarity of consumer-level multiprocessor hardware. *Although threaded* *applications incur little additional performance penalty on single-processor machines, the extra overhead of development has been difficult to justify due to the **preponderance** of single-processor machines.*

As of September 2006, with the typical mix of mass-market applications the main benefit to an ordinary user from a multi-core CPU will be improved on multitasking performance, which may apply more often than expected. Ordinary users have already been running many threads; operating systems utilize multiple threads, as well as antivirus programs and other "background processes" including audio and video controls. The largest boost in performance will likely be noticed in improved response time while running CPU-intensive processes, like antivirus scans, **defragmenting**, **ripping**/burning media (requiring file conversion), or searching for folders. For example, if the automatic virus scan initiates while a movie is being watched, the movie is far less likely to lag, as the antivirus program will be assigned to a different processor than the processor running the movie playback.

Given the increasing emphasis on multi-core chip design, stemming from the grave thermal and power consumption problems **posed** by any further significant increase in processor clock speeds, the extent to which software can be multithreaded to take advantage of these new chips is likely to be the single greatest constraint on computer performance in the future. If developers are unable to design software to fully exploit the resources provided by multiple cores, then they will ultimately reach an **insurmountable** performance ceiling.

The telecommunications market had been one of the first that needed a new design of parallel datapath packet processing because there was a very quick adoption of these multiple core processors for the datapath and the control plane. These **MPUs** are going to replace the traditional network processors that were based on proprietary micro- or **pico**-code. 6WIND was the first company to provide embedded software for these applications.

Parallel programming techniques can benefit from multiple cores directly. Some existing parallel programming models such as Cilk++, OpenMP and MPI can be used on multi-core platforms. Intel introduced a new **abstraction** for C++ parallelism called **TBB**. Other research **efforts** include the Codeplay Sieve System, Cray's Chapel, Sun's Fortress, and IBM's X10.

On the other hand, on the server side, multi-core processors are ideal because they allow many users to connect to a site simultaneously and have independent threads of execution. This allows for Web servers and application servers that have much better **throughput**.

◉ Licensing

Typically, proprietary enterprise server software is licensed "per processor". In the past a CPU was a processor and most computers had only one CPU, so there was no ambiguity.

Now there is the possibility of counting cores as processors and charging a customer for multiple licenses for a multi-core CPU. However, the trend seems to be counting dual-core chips as a single processor as Microsoft, Intel, and AMD support this view. Microsoft has said they would treat a socket as a single processor.

Oracle counts an AMD or Intel dual-core CPU as a single processor but has other numbers for other types, especially for processors with more than two cores. IBM and HP count a multi-chip module as multiple processors. If multi-chip modules count as one processor, CPU makers have an incentive to make large expensive multi-chip modules so their customers save on software licensing. So it seems that the industry is slowly heading towards counting each die as a processor, no matter how many cores each die has. Intel has released Paxville which is really a multi-chip module but Intel is calling it a dual-core - because it uses only one socket. It is not clear yet how licensing will work for Paxville. This is an unresolved and *thorny* issue for software companies and customers of *proprietary software*, leading many to consider *open source* alternative.

 ## Words

die[dai] *n.*管芯	**process**[prə'ses] *n.*进程
quad-[kwɔd] *adj.*由四部分组成的，四重的	**virus**['vaiərəs] *n.*病毒
monolithic[ˌmɔnə'liθik] *n.*单片电路，单块集成电路	**bug**[bʌg] *n.*程序缺陷，错误
cache[kæʃ] *n.*高速缓存	**interleaving**[ˌintə(:)'li:viŋ] *n.*交叉，交错
superscalar['sju:pə'skeilə] *n.*超标量体系结构	**debug**[di:'bʌg] *v.*调试
pipelining['paipˌlainiŋ] *n.*流水线操作	**break**[breik] *v.*暂停工作
multithreading['mʌlti'θrediŋ] *n.*多线程	**preponderance**[pri'pɔndərəns] *n.*优势
thread[θred] *n.*线程	**defragmenting**[di:'frægməntiŋ] *n.*磁盘碎片整理
algorithm['ælgəriðəm] *n.*算法	程序
speedup['spi:dʌp] *n.*加速	**rip**[rip] *v.*用程序将（激光唱盘上的音序）存储到
multitask['mʌltiˌtɑ:sk] *v.*做多重工作	硬盘上
discrepancy[dis'krepənsi] *n.*差异，矛盾	**pose**[pəuz] *v.*提出，造成
semantics[si'mæntiks] *n.*语义	**insurmountable**[ˌinsə'mauntəbl] *adj.*不能克服
proximity[prɔk'simiti] *n.*接近，亲近	的，不能超越的
geometry[dʒi'ɔmitri] *n.*几何，几何形状	**pico-**[ˈpaikəu] *adj.*兆分之一
yield[ji:ld] *n.*产量，收益	**abstraction**[æb'strækʃən] *n.*抽象
counter['kauntə] *v.*反对，辩驳	**effort**['efət] *n.*成果，努力的结果
commitment[kə'mitmənt] *n.*托付，交托，致力	**throughput**['θru:put] *n.*吞吐量
bandwidth['bændwidθ] *n.*带宽	**thorny**['θɔ:ni] *adj.*棘手的，麻烦的
heterogeneous[ˌhetərəu'dʒi:niəs] *adj.* 异构的	

Phrases

cache snoop	高速缓存监听	**portable media player**	便携式媒体播放器
bus snoop	总线监听	**proprietary software**	专有软件
count as	认为…，当作…	**open source**	开源
fine-grain	（影像）有微粒的，细致的		

Abbreviations

CMP	Chip-level MultiProcessor	芯片级多处理器
DSP	Digital Signal Processors	数字信号处理器
GPU	Graphics Processing Unit	图形处理单元
SoC	System-on-a-Chip	单片系统
SMP	Symmetric MultiProcessing	对称多处理
OS	Operating System	操作系统
DVFS	Dynamic Voltage and Frequency Scaling	动态电压与频率调节
MPU	MicroProcessor Unit	微处理器
TBB	Thread Building Blocks	线程构建模块

Complex Sentences

1. **Original:** The proximity of multiple CPU cores on the same die allows the cache coherency circuitry to operate at a much higher clock rate than is possible if the signals have to travel off-chip.

   Translation: 由于位于同一芯管上的几个CPU距离很近，高速缓存的同步要比在多个芯管间CPU上的高速缓存同步快很多。

2. **Original:** Although threaded applications incur little additional performance penalty on single-processor machines, the extra overhead of development has been difficult to justify due to the preponderance of single-processor machines.

   Translation: 虽然在单处理器机上线程应用程序几乎不会导致额外的性能损失，但是由于单处理器机的优势，开发的额外开支已经很难证明。

Exercises

I. Read the following statements carefully, and decide whether they are true (T) or false (F) according to the text.

_____1. A new wider core design will be a general trend in processor development.

_____2. In multi-core chip design, increase in processor clock speed may result in the grave thermal and power consumption problems.

_____3. This article uses both the terms "multi-core" and "dual-core" to refer to multiple physically separate CPU.

___4. It seems acceptable for the industry to count cores as processors and charge a customer for multiple licenses for a multi-core CPU.

___5. The technology of multi-core processors is widely used in other technology areas besides personal computer.

II. Choose the best answer to each of the following questions.

1. Which of the following is RIGHT about the multi-core processors?

(A) Amdahl's law indicates that the amount of performance gained by the use of a multi-core processor is not only dependent on their implementation in hardware itself.

(B) Multi-core architecture is able to balance the work evenly across multiple cores for each application.

(C) Cores in a multi-core device share a single coherent cache and the same interconnect to the rest of the system in all of the current dual-core processors.

(D) A system with n cores is always more effective than the ones that have only one core.

2. Which of the following is WRONG about the advantages of the multi-core CPU over multi-coupled single-core processors?

(A) Less PCB space occupying

(B) Less power consuming

(C) Higher performance of cache snoop operations

(D) More often repeating individual signals

3. Which of the following is NOT the constraint for the multi-core processor hardware design on increasing system performance according to this article?

(A) Production yields down

(B) Difficulty in thermal management

(C) Advanced architecture

(D) Shared system bus and memory bandwidth limits

III. Translating.

1. This all assumes however that the software has been designed to take advantage of available parallelism. If it hasn't, there will not be any speedup at all.

2. It would be possible for an application that used two CPUs to end up running faster on one dual-core if communication between the CPUs was the limiting factor, which would count as more than 100% improvement.

Part 2　**Simulated Writing: Notices**

◉Introduction

Notice is a kind of written news, announcement or information. People write notices for many reasons. No matter which kind of notice you need to write, you will find the following tips and format useful. Remember to tailor the tips to the specific notice you need to write, and get on your way.

➜ Tips on How to Write a Notice

- Be direct and concise in your notice. Your reader will be able to understand the information quickly and can refer back to it easily.
- Write a short, friendly notice that's to the point when you're sharing positive news. Written in the right tone, a notice can show a wish to keep up a business or personal relationship. It can also build on positive feelings like confidence, allegiance, and helpfulness.
- Recognize what others have achieved in your notice, and motivate your reader to reach similar goals.
- Present your information in a plain and complete way, so your reader will understand you the first time (and not ask questions later).
- If the news you are announcing is bad, write it in a direct statement. Add a message of understanding and optimism to your notice, in a respectful tone.

➜ Format

The following things must be kept in mind while writing a notice:
- Notices must be placed in a box.
- Notices must carry a heading.
- Date, time, venue and the activity must be written properly.
- Name of the writer and his post should be mentioned at the end of the notices.
- The body of the notices must be concise.

➜ Sample

NOTICES

ANNUAL DAY CELEBRATION

September 14, 2007

Attention all the students!

This is to inform you that our school is going to celebrate its ANNUAL DAY in the school hall, at 10:30 a.m. on Monday, October 16, 2007. The state education minister will preside over the program. Students who want to participate can give your names to the undersigned within a week.

Kelly Zhang
Representative of Student Union

Part 3   Listening and Speaking

>> Dialogue: *Referring to* Websites or Online *Forum* for Microsoft Developer

(Today is Monday. Henry and Mark are on the way to the classroom when they come across Sophie who is going to the classroom too.)

Henry & Mark	Good morning, Sophie.
Sophie	Good morning, Henry and Mark.
Henry	How's your weekend going, Sophie?
Sophie	I'm feeling very tired.
Mark	What's wrong?
Sophie	I have some problems with my C++[1] homework, which make me really confused.
Henry	What can we do for you?
Sophie	I know only a little about the class libraries in C++, and don't know how to use proper class libraries to accomplish specific functions in my applications.
Henry	Maybe you can find some valuable information on the Internet.
Sophie	Where can I find relevant information efficiently?
Mark	You can find answers on the Microsoft Developer Network, shorted by **MSDN**, a resource *Website* for developers offered by Microsoft[2]. Its URL is http://msdn.microsoft.com/.
Henry	Especially, the MSDN Library is an essential source of information for developers using Microsoft tools, products, technologies and services. The MSDN Library includes *how-to* and reference documentation, sample code, technical articles, and more. You can browse the table of contents or use search to find the content you need easily.[2]
Mark	In addition, I suggest you visit some software developer resource Websites provided by the third parties which offer large numbers of practical **cases** and corresponding solutions.

Henry	That's right! For example, The Code Project is a Website that contains articles for an audience of primary Microsoft computer programmers. Articles can be related to general programming, GUI design, algorithms or collaboration. Most of the articles have been uploaded by visitors and have not been extracted from an external source. Nearly every article is accompanied with source code and examples which can be downloaded independently. The Code Project contains articles and code pertaining to various programming languages: C/C++ , C#, Visual Basic, ASP, AJAX, SQL, and so on.[3]
Mark	Yes. Once you have registered a user account on The Code Project, you may write and upload your own articles and code for other visitors to view.
Sophie	When I have a question that has not been solved properly, can I ask other developers through these Websites?
Mark	Yes. There are a lot of developer forums and communities where you can share huge resources and experience with others.
Henry	You can visit and join some developer forum Websites, such as Source Forge and Experts-exchange. The Code Project also has fairly active forums, and is a reasonably good resource for resolving difficult software development issues.
Sophie	It's very kind of you to do me such a favor. Thank you vey much!
Henry & Mark	It's my pleasure.

Exercises

Work in pairs, and make up a similar conversation by replacing the scenarios with other material in the below.

Referring to Websites or Online Forum for Java Developer

[1] Java

[2] Sun Developer Network (**SDN**) is a Community for Sun Developers. It is a large and diverse community with both online and offline resources that can help you *address* and solve the technical issues you face everyday. The Sun Developer's core competency lies in the ability to connect development peers and supply timely access to the services that support the entire development lifecycle. Its URL is: http://developers.sun.com/.

[3] JavaWorld was launched in 1996 as the original independent resource for Java developers, architects, and managers. With a reputation for quality and a loyal following of up to one million visitors per month, JavaWorld's mission is to **mentor** and **inspire** its **readership** by publishing solutions-oriented content that is practical, timely, and **engaging**.

 Words

forum[ˈfɔːrəm] *n.*论坛	**inspire**[inˈspaiə] *v.*鼓舞，唤起
case[keis] *n.*案例，实例	**readership**[ˈriːdəʃip] *n.*（报刊、书等拥有的）读者（数）
address[əˈdres] *v.*处理，对付	**engaging**[inˈgeidʒiŋ] *adj.*有吸引力的，吸引人的
mentor[ˈmentɔː] *v.*指导	

 Phrases

refer to	查阅，参考	**come across**	偶然遇见
Website	网站	**how-to**	解释作法的，指南的

 Abbreviations

MSDN	Microsoft Developer Network	微软开发者网络
SDN	Sun Developer Network	Sun开发者网络

≫ Listening Comprehension: Intel

Listen to the passage and the following 3 questions based on it. After you hear a question, there will be a break of 10 seconds. During the break, you will decide which one is the best answer among the four choices marked (A), (B), (C) and (D).

Questions

1. How many persons with status in the history of Intel are mentioned in this article?
2. When did Intel go through the period of high-growth?
3. Which of the following factors promoted the wide adoption of Pentium processor in PC industry according to this article?

Choices

1. (A) One   (B) Two   (C) Three   (D) Four
2. (A) 1970s   (B) 1980s   (C) 1990s   (D) 2000s
3. (A) Speed processing ability   (B) Low price
   (C) Limited heat generation   (D) Successful advertising logo

A Words

headquarter[ˌhedˈkwɔːtə] v.以…作总部，设总公司于…	**unprecedented**[ʌnˈpresidəntid] adj.空前的，史无前例的
giant[ˈdʒaiənt] n.巨人，伟人	**stumble**[ˈstʌmbl] v.跌绊
assert[əˈsəːt] v.断言，声称	**engage**[inˈgeidʒ] v.使卷入其中，与…交战
premier[ˈpremjə] adj.首要的，第一的	**bruise**[bruːz] v.撞伤，碰伤
embark[imˈbɑːk] v.开始，从事	

≫ Dictation: Father of the Mouse — Doug Engelbart

This passage will be played THREE times. Listen carefully, and fill in the blanks with the words you have heard.

Many people ____1____ believe that the ***mouse*** was invented by Apple. Others believe that Steve Jobs ____2____ the idea from Xerox, where the mouse was used on an early office PC called the Star. But ____3____ , the mouse was first ***conceived*** by Doug Engelbart in the early 1960's, then a scientist at the Stanford Research Institute, in Menlo Park, California.

As a pioneer of human-computer ____4____ , Douglas Engelbart changed the way computers worked, from specialized machinery that only a trained scientist could use, to a user-friendly ____5____ that almost anyone can use. He invented or ____6____ to several interactive, user-friendly devices: the computer mouse, windows, computer video ____7____ , ***hypermedia***, ***groupware***, email, the Internet and more. In 1964, the first ***prototype*** computer mouse was made to use with a graphical user interface, "windows". Engelbart received a ***patent*** for the wooden ____8____ with two metal ____9____ in 1970, describing it in the patent ____10____ as an "X-Y position ***indicator*** for a display system." "It was ____11____ the mouse because the ____12____ came out in the end," Engelbart ***revealed*** about his invention.

Engelbart has always been ____13____ of his time, having ideas that seemed ***far-fetched*** at the time but later were taken for granted. Besides the computer mouse, one of his most famous inventions in 1960s, not used ____14____ until the 1980s, the huge success of Microsoft's Windows 95 ____15____ that Engelbart's original windows ____16____ in 1950 has also become a virtual necessity.

Engelbart wanted to use technology to ***augment*** ____17____ . He saw technology, especially computers, as the answers to the problem of ____18____ the ever more complex modern world and has ***dedicated*** his life to the ***pursuit*** of developing technology to augment human intellect. He seldom receives the ____19____ many believe he ***deserves***. In 1989, he founded the Bootstrap

Institute to *foster* high performance organizations by developing enabling technologies and promoting _____20_____ . At the age of 83 now, his work continues.

 Words

mouse[maus] *n.*鼠标	**reveal**[ri'vi:l] *v.*展现，显示
conceive[kən'si:v] *v.*构思，设计	**far-fetched**['fɑː'fetʃt] *adj.*牵强的，不自然的
hypermedia['haipəmi:diə] *n.*超媒体	**augment**[ɔː g'ment] *v.*扩大，增加
groupware[gru:pwɛə] *n.*组件，群件	**dedicate**['dedikeit] *v.*奉献，致力
prototype['prəutətaip] *n.*原型	**pursuit**[pə'sju:t] *n.*追求，追寻
patent['peitənt] *n.*专利权	**deserve**[di'zə:v] *v.*应受，应得
indicator['indikeitə] *n.*指针，指示器	**foster**['fɔstə] *v.*促进，培养

Computer Software

Unit
3

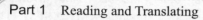

Part 1   Reading and Translating

⫸ Section A   Microsoft Visual Studio

*Microsoft Visual Studio is the main Integrated Development Environment (**IDE**) from Microsoft. It can be used to develop console and Graphical User Interface applications along with Windows **Forms** applications, Web sites, Web applications, and Web services in both native code as well as managed code for all platforms supported by Microsoft Windows, Windows Mobile, .NET Framework, .NET Compact Framework and Microsoft Silverlight.*

Visual Studio includes a code editor supporting *IntelliSense* as well as code ***refactoring***. The integrated debugger works both as a source-level debugger and a machine-level debugger. Other ***built-in*** tools include a forms designer for building GUI applications, Web designer, ***class*** designer, and ***database schema*** designer. It allows ***plug-in***s to be added that enhance the functionality at almost every level – including adding support for source control systems (like Subversion and Visual SourceSafe) and adding new toolsets like editors and visual designers for domain-specific languages or toolsets for other aspects of the software development lifecycle (like the Team Foundation Server client: Team Explorer).

Visual Studio supports languages by means of language services, which allow any programming language to be supported (to varying degrees) by the code editor and debugger, provided a language-specific service has been **authored**. Built-in languages include C/C++ (via Visual C++), VB.NET (via Visual Basic .NET), and C# (via Visual C#). *Support for other languages such as Chrome, F#, Python, and Ruby among others has been made available via language services which are to be installed separately.* It also supports **XML/XSLT**, **HTML/ XHTML**, JavaScript and **CSS**. Language-specific versions of Visual Studio also exist which provide more limited language services to the user. These individual packages are called Microsoft Visual Basic, Visual J#, Visual C#, and Visual C++.

Currently, Visual Studio 2008 and 2005 Professional Editions, along with language-specific versions (Visual Basic, C++, C#, J#) of Visual Studio 2005 are available to students as downloads *free of charge* via Microsoft's DreamSpark program.

Visual Studio 2008

Visual Studio 2008, **codenamed** Orcas, is the successor to Visual Studio 2005. It was released to MSDN subscribers on 19 November 2007 alongside .NET Framework 3.5. The codename Orcas is, like Whidbey, a reference to an island in Puget Sound, Orcas Island. The source code for the Visual Studio 2008 IDE will be available under a shared source license to some of Microsoft's partners and **ISVs**. Microsoft released Service Pack 1 for Visual Studio 2008 on 11 August 2008.

Visual Studio 2008 is focused on development of Windows Vista, 2007 Office system, and Web applications. Among other things, it brings a new language feature, **LINQ**, new versions of C# and Visual Basic languages, a **Windows Presentation Foundation** visual designer, and improvements to the .NET Framework. It also features a new HTML/CSS editor influenced by Microsoft Expression Web. J# is not included. Visual Studio 2008 requires .NET Framework 3.5 and by default configures compiled assemblies to run on .NET Framework 3.5; but it also supports multi-targeting which lets the developers choose which version of the .NET Framework (out of 2.0, 3.0, 3.5, Silverlight CoreCLR or .NET Compact Framework) the assembly runs on. Visual Studio 2008 also includes new code analysis tools, including the new Code Metrics tool. For Visual C++, Visual Studio adds a new version of Microsoft Foundation Classes (**MFC** 9.0) that adds support for the visual styles and UI controls introduced with Windows Vista. For native and managed code interoperability, Visual C++ introduces the **STL/CLR**, which is a port of the C++ Standard Template Library (STL) containers and algorithms to managed code. STL/CLR defines STL-like containers, *iterators* and algorithms that work on C++/**CLI** managed objects.

Visual Studio 2008 features a **XAML** based designer (codenamed Cider), *workflow* designer, LINQ to **SQL** designer (for defining the type mappings and *object encapsulation* for SQL Server data), XSLT debugger, JavaScript Intellisense support, JavaScript Debugging support, support for **UAC** *manifests*, a *concurrent* build system, among others. It ships with an enhanced set of UI *widgets*, both for WinForms and WPF. *It also includes a multithreaded build engine (MSBuild) to compile multiple source files (and build the executable file) in a project across multiple threads simultaneously.* It also includes support for compiling **PNG** compressed *icon* resources introduced in Windows Vista. An updated XML Schema designer will ship separately some time after the release of Visual Studio 2008.

The Visual Studio debugger includes features targeting easier debugging of multi-threaded applications. In debugging mode, in the threads window, which lists all the threads, hovering over a thread will display the *stack* trace of that thread in *tooltips*. The threads can directly be named and flagged for easier identification from that window itself. In addition, in the code window, along with the location information of the currently executing instruction in the current thread, the currently executing instructions in other threads are also pointed out. The Visual Studio debugger supports integrated debugging of the .NET Framework 3.5 **BCL**. It can dynamically download the BCL source code and debug symbols and allow stepping into the BCL source during debugging. Currently a limited subset of the BCL source is available, with more library support planned for later in the year.

Future Development

A future iteration of Visual Studio, codenamed Visual Studio 10 or VS 10 (previously codenamed Hawaii), is under development. Even though the feature set has not been finalized, some features that the teams are considering have been made public. The Visual C++ team is considering using a SQL Server Compact database to store information about the source code, including IntelliSense information, for better IntelliSense and code-completion support. For managed code, a Call Hierarchy feature, which will show all the code-paths from any method at design time, is being designed. The Visual Studio 10 IDE is also *slated* to be redesigned to be more modular and with more extensibility points than the current version.

Visual Studio Rosario is the next version of Visual Studio Team System and is being *touted* as an "integrated Application Life-cycle Management (**ALM**)" tool. It is intended to enable and enhance development at every step of an application's life-cycle from conceptualization to release and maintenance. Microsoft is currently *soliciting* comments and feedback from developers for Rosario.

Words

forms[fɔ:mz] *n.*窗体	**object**[ˈɔbdʒikt] *n.*对象
refactor[ˈri: ˈfæktə] *v.*重构	**encapsulation**[inˌkæpsjuˈleiʃ ən] *n.*封装
built-in[ˈbiltˈin] *adj.*内置的	**manifest**[ˈmænifest] *n.*显示，清单
class[klɑ:s] *n.*类	**concurrent**[kənˈkʌrənt] *adj.*并发的
database[ˈdeitəbeis] *n.*数据库	**widget**[ˈwidʒit] *n.*窗口小部件，小工具
schema[ˈski:mə] *n.*模式	**icon**[ˈaikɔn] *n.*图标
plug-in[plʌgˈin] *n.*插件程序	**stack**[stæk] *n.*栈
author[ˈɔ:θə] *v.*创造	**slate**[sleit] *v.*安排，指定
code-name[kəudˈneim] *v.*给与代号	**tout**[taut] *v.*吹捧
iterator[ˈitəreitə] *n.*迭代器	**solicit**[səˈlisit] *v.*恳求获得
workflow[ˈwə:kfləu] *n.*工作流	

 Phrases

IntelliSense	智能提示	**free of charge**	免费
among others（other things）	其中	**tooltips**	工具提示

 Abbreviations

IDE	Integrated Development Environment	集成开发环境
XML	eXtensible Markup Language	可扩展标记语言
XSL	eXtensible Stylesheet Language	可扩展样式表语言
XSLT	XSL Transformations	XSL转换
HTML	HyperText Markup Language	超文本标记语言
XHTML	eXtensible HyperText Markup Language	可扩展超文本标记语言
CSS	Cascading Style Sheets	层叠样式表
ISV	Independent Software Vendors	独立软件开发商
LINQ	Language INtegrated Query	语言级集成查询
WPF	Windows Presentation Foundation	微软用于Windows的统一显示子系统
MFC	Microsoft Foundation Classes	微软基础类
STL	Standard Template Library	标准模板库
CLR	Common Language Runtime	公用语言运行时
CLI	Common Language Infrastructure	通用语言基础结构
XAML	eXtensible Application Markup Language	可扩展应用程序标记语言
SQL	Structured Query Language	结构化查询语言
UAC	User Account Control	用户账户控制
PNG	Portable Network Graphic	可移植的网络图像文件格式
BCL	Base Class Library	基类库
ALM	Application Life-cycle Management	应用程序生命周期管理

■ Complex Sentences

1. **Original:** Visual Studio supports languages by means of language services, which allow any programming language to be supported (to varying degrees) by the code editor and debugger, provided a language-specific service has been authored.

   Translation: 如果创建了特定的语言服务，Visual Studio就可以用语言服务来支持语言，这个语言服务允许代码编辑器和调试器（在不同程度上）支持任何一种编程语言。

2. **Original:** Visual Studio 2008 also includes a multithreaded build engine (MSBuild) to compile multiple source files (and build the executable file) in a project across multiple threads simultaneously.

   Translation: Visual Studio 2008也包含一个多线程构建引擎（MSBuild），这个引擎可以在一个工程中同时跨多个线程编译多个源文件（并创建可执行文件）。

■ Exercises

I. Read the following statements carefully, and decide whether they are true (T) or false (F) according to the text.

____1. Visual Studio has six built-in languages including C/C++, Visual C++, VB.NET, Visual Basic .NET, C# and Visual C#.

____2. The integrated debugger is a plug-in for Visual Studio to enhance the debugging function at source-level and machine-level.

____3. Visual Studio 2008 is used to develop Windows Vista and 2007 Office system.

____4. The feature set of Visual Studio 10 has been made public.

____5. Language-specific versions of Visual Studio provide four individual packages (Microsoft Visual Basic, Visual J#, Visual C#, and Visual C++) as language services.

II. Choose the best answer to each of the following questions.

1. Which statement is RIGHT about Visual Studio 2008?

   (A) Visual Studio MFC 9.0 can add support for the visual styles and UI controls introduced with Windows Vista.

   (B) Compiled assemblies configured by Visual Studio 2008 needs to run on .NET Framework 3.5.

   (C) The source code for the Visual Studio 2008 IDE is available to students as downloads free of charge via Microsoft's DreamSpark program.

   (D) STL/CLR is a specific version of the C++ STL containers and algorithms that work on C++/CLI managed objects.

2. What is the main idea of the last paragraph in the section "Visual Studio 2008"?

   (A) The Visual Studio debugger includes features targeting easier debugging of multi-threaded applications.

(B) The Visual Studio debugger can operate synchronously in both the threads window and the code window in debugging mode.

(C) The Visual Studio debugger can list all threads, display the stack trace of a hovered thread, and indicate the location of the currently executing instructions in the current thread and other threads synchronously.

(D) The Visual Studio debugger supports integrated debugging of the .NET Framework 3.5 BCL.

3. Which of the following code name does NOT belong to the same product series as other three ones?

(A) Whidbey   (B) Orcas   (C) Hawaii   (D) Rosario

III. Translating.

1. Microsoft Visual Studio can be used to develop console and Graphical User Interface applications along with Windows Forms applications, Websites, Web applications, and Web services in both native code as well as managed code for all platforms supported by Microsoft Windows, Windows Mobile, .NET Framework, .NET Compact Framework and Microsoft Silverlight.

2. Support for other languages such as Chrome, F#, Python, and Ruby among others has been made available via language services which are to be installed separately.

≫ Section B   Google Earth

Google Earth is a proprietary virtual globe program that was originally called Earth Viewer, and was created by Keyhole Inc, a company acquired by Google in 2004. *It maps the earth by the* ***superimposition*** *of images obtained from satellite **imagery**, **aerial** photography and* **GIS 3D** *globe.*

The product, renamed Google Earth in 2006, is currently available for use on personal computers running Microsoft Windows 2000, XP, or Vista, Mac OS X 10.3.9 and above, Linux (released on June 12, 2006), and FreeBSD. Google Earth is also available as a ***browser*** plug-in (released on June 02, 2008) for Firefox, IE6, or IE7. In addition to releasing an updated Keyhole based client, Google also added the imagery from the Earth database to their Web based mapping software. The release of Google Earth in mid 2006 to the public caused a more than tenfold increase in media coverage on virtual globes between 2006 and 2007, driving public interest in geospatial technologies and applications.

Google Earth displays satellite images of varying ***resolution*** of the Earth's surface, allowing users to visually see things like houses and cars from a bird's eye view. The degree of resolution available is based somewhat on the points of interest, but most land (except for some islands) is covered in at least 15 meters of resolution. Melbourne, Australia, Las Vegas, Nevada and

Cambridge, Massachusetts include examples of the highest resolution, at 15 cm (6 inches). Google Earth allows users to search for addresses for some countries, enter **coordinates**, or simply use the mouse to browse to a location.

Google Earth also uses digital elevation model (**DEM**) data collected by **NASA**'s Shuttle Radar Topography Mission (**SRTM**). This means one can view the *Grand Canyon* or *Mount Everest* in three dimensions, instead of 2D like other map programs/sites. Since November 2006, the 3D views of many mountains, including Mount Everest, have been improved by the use of supplementary DEM data to fill the gaps in SRTM coverage.

Many people using the applications are adding their own data and making them available through various sources, such as the **BBS** or **blogs**. Google Earth is able to show all kinds of images overlaid on the surface of the earth and is also a **Web Map Service** client. Google Earth supports managing three-dimensional geospatial data through Keyhole Markup Language (**KML**).

Google Earth has the capability to show 3D buildings and structures (such as bridges), which consist of users' submissions using SketchUp, a 3D modeling program. In prior versions of Google Earth (before Version 4), 3D buildings were limited to a few cities, and had poorer **rendering** with no **textures**. Many buildings and structures from around the world now have detailed 3D structures; including (but not limited to) those in the United States, Canada, Ireland, India, Japan, United Kingdom, Germany, Pakistan and the cities, Amsterdam and Alexandria. In August 2007, Hamburg became the first city entirely shown in 3D, including textures such as **facades**. The Irish town of Westport was added to Google Earth in 3D on January 16th, 2008. The "Westport3D" model was created by 3D imaging firm AM3TD using long-distance laser scanning technology and digital photography and is the first such model of an Irish town to be created. As it was developed initially to aid local government in carrying out their town planning functions it includes the highest resolution photo-realistic textures to be found anywhere in Google Earth. Three-dimensional renderings are available for certain buildings and structures around the world via Google's 3D Warehouse and other Websites.

Recently, Google added a feature that allows users to monitor traffic speeds at loops located every 200 yards in real-time.

In version 4.3 released on April 15, 2008, Google Street View was fully integrated into the program allowing the program to provide an on the street level view in many locations.

National Security and Privacy Issues

The software has been criticized by a number of special interest groups, including national officials, as being an invasion of privacy and even posing a threat to national security. The typical argument is that the software provides information about military or other critical installations that could be used by terrorists.

Some citizens may express concerns over aerial information depicting their properties and residences being **disseminated** *freely. As relatively few* **jurisdictions** *actually guarantee the individual's right to privacy, as opposed to the state's right to secrecy, this is an evolving, but minor, point.* Perhaps aware of these critiques, *for a time,* Google had Area 51 (which is highly visible and easy to find) in Nevada as a default place mark when Google Earth is first installed.

As a result of pressure from the United States government, the residence of the Vice President at Number One Observatory Circle is obscured through *pixelization* in Google Earth and Google Maps. The usefulness of this *downgrade* is questionable, as high-resolution photos and aerial surveys of the property are readily available on the Internet elsewhere. *Capitol Hill* used to also be pixelized in this way but this was *lifted*.

Critics have expressed concern over the willingness of Google to **cripple** *their dataset to* **cater** *to special interests, believing that intentionally obscuring any land goes against its stated goal of letting the user "point and* **zoom** *to any place on the planet that you want to explore."*

Finally, **empirical** *research has shown that while Google does allow people to* opt-out *from personal listings, a vast majority of people can still be geographically located using phone numbers.*

 Words

superimposition[ˈsjuːpərˌimpəˈziʃən] *n.*叠印	**disseminate**[diˈsemineit] *v.*散布
imagery[ˈimidʒəri] *n.*影像	**jurisdiction**[ˌdʒuərisˈdikʃən] *n.*司法，权限
aerial[ˈɛəriəl] *adj.*航空的，由飞机进行的	**pixelization**[piksəlaiˈzeiʃən] *v.*像素化
browser[brauzə] *n.*浏览器	**downgrade**[ˈdaungreid] *n.*向下渐变
resolution[ˌrezəˈljuːʃən] *n.*分辨率	**lift**[lift] *v.*解除
coordinates[kəuˈɔːdinit] *n.*坐标	**cripple**[ˈkripl] *v.*削弱
blog[blɔg] *n.*博客	**cater**[ˈkeitə] *v.*满足（需要），投合
rendering[ˈrendəriŋ] *n.*表现，渲染	**zoom**[zuːm] *v.*移向（或移离）目标
texture[ˈtekstʃə] *n.*纹理	**empirical**[emˈpirikəl] *adj.*完全根据经验的，经验
facade[fəˈsɑːd] *n.*（房屋的）正面，立面	主义的

 Phrases

Grand Canyon	（美）科罗拉多大峡谷	**Capitol Hill**	美国国会山，美国国会
Mount Everest	珠穆朗玛峰	**opt-out**	宣布放弃选择（权）
for a time	暂时		

Abbreviations

GIS	Geographic Information System	地理信息系统
3D	Three-Dimensional	三维的
DEM	Digital Elevation Model	数字高程模型
NASA	National Aeronautics and Space Administration	（美国）国家航空航天局
SRTM	Shuttle Radar Topography Mission	航天飞机雷达地形测绘任务
BBS	Bulletin Board System	电子公告栏系统
WMS	Web Map Service	网络地图服务
KML	Keyhole Markup Language	Keyhole标记语言，是一个基于XML语法和文件格式的文件，用来描述和保存地理信息如点、线、图片、折线并在Google Earth客户端之中显示

Complex Sentences

1. **Original:** Some citizens may express concerns over aerial information depicting their properties and residences being disseminated freely. As relatively few jurisdictions actually guarantee the individual's right to privacy, as opposed to the state's right to secrecy, this is an evolving, but minor, point.

   Translation: 一些公民也许对那些正在被随意传播的描述其财产和住处的航空拍摄信息表示关注。当与国家保密的权利相比时，实际上只有相对很少的司法来保证个人隐私的权利，虽然这种观点在不断被人们接受，但仍显得微不足道。

2. **Original:** Critics have expressed concern over the willingness of Google to cripple their dataset to cater to special interests, believing that intentionally obscuring any land goes against its stated goal of letting the user "point and zoom to any place on the planet that you want to explore."

   Translation: 批评者已经对Google公司愿意削减数据集来满足特定利益的做法表示关注，认为故意地模糊化任何国土违反了其宣称的目标，即让用户"指向并移动到用户想探查的地球上的任何地方"。

Exercises

I. Read the following statements carefully, and decide whether they are true (T) or false (F) according to the text.

____1. Google earth is designed and created by Google as an initiator in 3D globe in 2004.

____2. The highest resolution on Google Earth is 6 inches per pixel.

____3. Google Earth was originally a browser plug-in before it's released as an updated Keyhole based client.

___4. Now, Google Earth can run on many kinds of operating system.

___5. Pixelizing the imagery of some specific place in Google Earth and Google Maps is proved to be an effective way to protect personal privacy and national security.

II. Choose the best answer to each of the following questions.

1. Which of the following is **WRONG** about the impacts of Google Earth to the public?

(A) Cause a more than tenfold increase in media coverage on virtual globes.

(B) Raise the concern over invasion of privacy and even posing a threat to national security.

(C) Drive public interest in geospatial technologies and applications.

(D) Replace the traditional geographical locating mode using phone numbers.

2. Which kind of data allows Google Earth to map natural terrain in three dimensions?

(A) Earth Viewer created by Keyhole   (B) DEM data collected by NASA's SRTM

(C) Westport3D data created by AM3TD   (D) KML data created by SketchUp

3. Which of the following is NOT mentioned about the uses of Google Earth in this article?

(A) Locate and browse a place on the planet   (B) Aid to town planning

(C) Sell images as a Web Map Service provider (D) Monitor traffic speed

III. Translating.

1. Google Earth maps the earth by the superimposition of images obtained from satellite imagery, aerial photography and GIS 3D globe.

2. Finally, empirical research has shown that while Google does allow people to opt-out from personal listings, a vast majority of people can still be geographically located using phone numbers.

Part 2   Simulated Writing: Report

➲Introduction

A report is a structured written presentation directed to interested readers in response to some specific purpose, aim or request. There are many varieties of reports, but generally their function is to give an account of something, to answer a question, or to offer a solution to a problem. An effective report is appropriate to its purpose and audience, accurate, logical, clear and concise, and well organized with clear section headings.

➲Report Structure

One important advantage that a report has over other written communication is that it follows a standardized format. This enables readers to find and focus on specific pieces of information. Most reports are modeled on the following structure (modified where necessary).

- Transmittal document
- Title page
- Table of contents
- Abstract/Executive Summary
- Introduction
- Discussion
- Conclusions
- Recommendations
- Bibliography

Procedure for Report Writing

The following is a suggestion as to how you might proceed in compiling and presenting a report. There are three stages.

Stage One: Planning

Stage Two: Writing

Stage Three: Formatting, revising and proof-reading

Stage One: Planning

1. Defining the purpose

Read the brief carefully, identify key words and make sure you know what's really being asked.

2. Defining the audience

Determine your audience's level of understanding and what your audience needs to know.

3. Establishing parameters

Determine the scope and level of detail required, the length of the report and what can be covered in that length.

4. Gathering information

Make sure the information you gather is relevant, contemporary and factually correct and make sure that you transcribe facts and figures correctly

Stage Two: Writing

1. Writing the body

There are four components of the body of the report: the introduction, the discussion, the conclusions and the recommendations.

(1) Introduction

The introduction leads into the main subject matter by giving the necessary background of the report, its aims, premises, scope, limitations, approach intended audience, possible benefits and any instructions that may be useful for the reader. If professional terms are used in the report, define them clearly.

It puts the discussion in perspective, explains why the report is necessary and gives background information on the subject matter.

(2) Discussion

The discussion is the main body of the report. Use headings and sub-headings. It describes, analyses, interprets and evaluates the procedures, data, findings, relationships, visual material, methodology and results in the report. This material should be presented in an order that leads logically towards the conclusions and recommendations.

In writing the discussion section of the body, you should:

- Pitch at appropriate level
- Organize material logically
- Use clear, concise language
- Give concrete examples

(3) Conclusions

Conclusions are drawn from evidence, analysis, interpretation and evaluation presented in the discussion. No new material should be introduced; the conclusions should follow logically from the Discussion. The Conclusions section should give:

- Conclusions
- Key points
- Main findings

(4) Recommendations

The Recommendations section (when used - not all reports give recommendations) should present your informed opinions, suggestions, possible actions to be taken, applications and recommendations arising from a rational consideration of the discussion and conclusions.

- Be definite
- Be perceptive

- Be imaginative
- Be rational

2. Abstract/Executive Summary

Once the body of the report is written, write the abstract. The abstract (also known as the executive summary) is a concise summary presentation of the essential elements of the report, from the introduction to the recommendations. It should be independent (can be read on its own), comprehensive (covers all the main points), clear and concise. As a general rule it should be short, only 10-15% of the length of the report, and should be written in full sentences and paragraphs. It should include a summary of the following:

- Purpose
- Scope
- Achievements
- Main points
- Conclusions
- Recommendations

3. Writing the supplementary material

(1) Transmittal document

The transmittal document is not part of the report, but accompanies the report. In letter, memo, or minute form, it personalizes the report for a specific reader and calls attention to those items or sections in the report which are of particular interest to that person.

(2) Title page

Identifies the report with the following information:

- Title
- Author's name, position and qualifications
- Authority for report
- Place of origin
- Date

(3) Table of contents

The table of contents shows the section titles and major headings in order of appearance, along with their related page locations. Standard page numbering begins with the Introduction. The Abstract or Executive Summary is usually numbered with lower case Roman numerals (i, ii, iii, iv, etc.)

(4) Bibliography

The bibliography lists all publications either cited or referred to in preparing the report. Use the Referencing System recommended by your school.

Stage Three: Formatting, revising and proof-reading

Apply the following "report checklist"

- Have I fulfilled the purpose of the report?
- Is it written at a level appropriate to its audience?
- Are its facts correct?
- Is it comprehensive?
- Is all the included information relevant?
- Are the layout and presentation well thought out?
- Is the style clear, concise and professional?
- Is the abstract compact?
- Does the introduction adequately introduce the discussion?
- Is the discussion organized logically?
- Does the conclusions section evaluate the discussion properly?
- Are the recommendations reasonable?
- Does the table of contents correspond with the actual contents? Are page numbers correct?
- Have I acknowledged all sources of information through correct referencing?
- Have I checked spelling, grammar and punctuation?
- Have I carefully proof-read the final draft?

Sample

Report on the Possibilities of the New Project

Terms of reference

The Board of Directors has requested a report on what our staff think of our plan of launching a new project in the second half year. They want to know the possibility of the plan. The report should be discussed for the next Board of Directors' meeting.

Proceedings

I read and analyzed the written comments made by nearly 200 staff members.

Findings

1. The case for the new project

1.1 Almost 50 percent of staff think the new project is badly needed to challenge the competitors of the company and help the company out of the crisis.

1.2 Nearly a quarter of staff have lost their heart about the old pattern of production and welcome a better improved project.

2. The case against the new project

2.1 About 8 percent of staff suspect the future of the new project.

2.2 About 8 percent were worried about the plan since it cost too much.

Conclusions

Most of our staff understand the importance of the new project.

The budget of the project causes a lot of worries.

Recommendations

The company should introduce the new project to the staff, emphasize its importance and do the best to cut down the expense of the project.

Robert Bridge
Marketing Manager

[The abstract/executive summary and the supplementary material are omitted from this report.]

Part 3 Listening and Speaking

Dialogue: Making an Electronic Album Using Multimedia Editing Software

(Today is the first day after the National Day holiday. Henry met Mark in the hall.)

Henry	Hi, Mark. How was your National Day holiday?
Mark	It's wonderful! During this holiday, I went to Hangzhou with my family. It's a very beautiful city. We took a lot of photos and made many pieces of video with my video camera.[1]
Henry	Really? Sounds exciting! All of these will be a precious memory. I think you can make a family album about your journey in Hangzhou with that material, so that you can enjoy it on your computer at any time. Further more, if you like, you can release it on your blog so as to allow more people to share with you.[2]

Mark	Oh, it's a good idea! But I don't know how to do that at all. Would you like to help me?
Henry	Sorry, I'm a **layman** too. But Sophie is good at multimedia editing software[3], maybe she can help you.
Sophie	According to my experience, graphics software is necessary, Mark. It can help you create, manipulate, and print graphics.
Henry	It has many types, right?
Sophie	Yes, it includes painting software, photo editing software, drawing software, 3-D graphics software, CAD software, and presentation software, etc.[4] However, in your case, Mark, photo editing software is enough, such as Photoshop[5].
Mark	Oh, yes. I've heard that it's a very nice photo editing software. And what about those pieces of video?
Sophie	Don't worry. You can edit those pieces of video with video editing software. It provides a set of tools for transferring video **footage** from a **camcorder** to a computer, clipping out unwanted footage, assembling video segment in any sequence, adding special visual effects and adding a sound track. [6]
Henry	I have heard that one brand of this software is Adobe Premiere.
Sophie	That's right! Besides, **DVD** authoring software offers tools for creating DVDs with Hollywood-style menus. For example, Sonic DVDit, ULead DVD MovieFactory, Apple iDVD, and Adobe Encore DVD.
Mark	Thank you very much for your helpful guide, Sophie. Would you like to tell me some details about how to use them to make a complete electronic album[7]?
Sophie	No problem! If you have time, I'll show you how to use these kinds of software.
Henry	Well, we'll expect your wonderful works, Mark!
Mark	Ok, I'll try my best!
Sophie	May you succeed!

 **Exercises**

Work in pairs, and make up a similar conversation by replacing the scenarios with other material in the below.

Making an Web Site Using Webpage Designing Software

[1] Mark joined in a drama union in the school and played a role in a drama. They've collected many wonderful pieces of *acts* in the video tapes.

[2] A drama union Website for presenting wonderful works and releasing the newest program notice of the union.

[3] Webpage designing software

[4] Macromedia Studio was a suite of programs designed for Web content creation distributed by Macromedia. Macromedia Studio 8 was the last version of Macromedia Studio. It comprised Dreamweaver 8, Flash 8, Flash 8 Video Converter, Fireworks 8, Contribute 3 and FlashPaper.

[5] Fireworks is used to design and create the content of the Web pages, especially for the image editing.

[6] Dreamweaver is used for the frame designing and the Website deployment.

[7] A simple but complete Website.

 Words

layman['leimən] *n.*外行	**camcorder**['kæmkɔːdə] *n.*可携式摄像机
footage['futidʒ] *n.*连续镜头，电影胶片	**act**[ækt] *n.*节目

 Abbreviations

DVD	Digital Video Disc	数字化视频光盘

>>> Listening Comprehension: The Software Giant — Microsoft

Listen to the passage and the following 3 questions based on it. After you hear a question, there will be a break of 10 seconds. During the break, you will decide which one is the best answer among the four choices marked (A), (B), (C) and (D).

Questions

1. What is the relationship between Paul Allen and Bill Gates?
2. Which product introduced a graphical user interface first?
3. Which item is NOT included in the characteristics of the "cloud" computing?

Choices

1. (A) Predecessor & Successor   (B) Employee & Employer
   (C) Competitors   (D) Partners & Schoolmates
2. (A) MS-DOS operating system   (B) Macintosh operating system
   (C) Windows operating system   (D) Linux operating system
3. (A) Connected via Windows   (B) Provided over the Internet
   (C) Free of charge   (D) Sold in a box

Words

explode[iks'pləud] *v.*猛增	**dominance**['dɔminəns] *n.*优势，统治
beneficiary[beni'fiʃəri] *n.*受惠者，受益人	**seamlessly**['si:mlisli] *adv.*无缝地，连续地

≫ Dictation: Embedded Systems

This passage will be played THREE times. Listen carefully, and fill in the blanks with the words you have heard.

Embedded system is a _____1_____ computer system that is _____2_____ a larger system or machine, designed for a particular kind of application device. An embedded system is some combination of computer _____3_____ , either fixed in _____4_____ or programmability.

In the earliest years of computers in the 1930-40s, computers were sometimes dedicated to a single task, but were far too large and expensive for most kinds of tasks _____5_____ by embedded computers of today. Over time however, the _____6_____ of programmable controllers evolved from _____7_____ electromechanical **sequencers**, via solid state devices, to the use of computer technology. One of the first _____8_____ modern embedded systems was the Apollo Guidance Computer, developed by Charles Stark Draper at the MIT Instrumentation Laboratory.

Certain operating systems or language platforms are **tailored** for the _____9_____, such as Embedded Java and Windows XP Embedded. However, some low-end consumer products use very inexpensive microprocessors and limited _____10_____ , with the application and operating system both part of a _____11_____ program. Typically, an embedded system is _____12_____ on a single microprocessor _____13_____ with the programs permanently _____14_____ in ROM, rather than being loaded into RAM, as programs on a personal computer are.

Embedded systems control many of the common devices in use today. ____15____ all *appliances* that have a digital interface — watches, microwaves, cameras, cars — ____16____ embedded systems. Since the embedded system is dedicated to specific tasks, design ____17____ can *optimize* it,____18____ the size and cost of the product, or increasing the *reliability* and ____19____. Some embedded systems are mass-produced, benefiting from economies of____20____.

 Words

sequencer[ˈsiːkwənsə] *n.*程序装置，定序器	**optimize**[ˈɔptimaiz] *v.*使最优化
tailor[ˈteilə] *v.*裁剪	**reliability**[riˌlaiəˈbiliti] *n.*可靠性
appliance[əˈplaiəns] *n.*用具，装置	

Operating System

Unit 4

Part 1　Reading and Translating

➤➤ Section A　Windows Vista

Windows Vista is a line of operating systems developed by Microsoft for personal computers, including home and business desktops, laptops, Tablet PCs, and media center PCs. Prior to its announcement on July 22, 2005, Windows Vista was known by its codename Longhorn. Development was completed on November 8, 2006; over the following three months it was released in stages to computer hardware and software manufacturers, business customers, and retail channels. On January 30, 2007, it was released worldwide, and was made available for purchase and download from Microsoft's Website. The release of Windows Vista came more than five years after the introduction of its predecessor, Windows XP, the longest time span between successive releases of Microsoft Windows.

Windows Vista contains many changes and new features, including an updated graphical user interface and visual style dubbed Windows Aero, improved searching features, new multimedia creation tools such as Windows DVD Maker, and redesigned networking, audio, print, and display sub-systems. Vista also aims to increase the level of communication between machines on a home network, using ***peer-to-peer*** technology to simplify sharing files and digital media between computers and devices. Windows Vista includes version 3.0 of the .NET Framework, which aims to make it significantly easier for software developers to write applications than with the traditional Windows **API.**

Microsoft's primary stated objective with Windows Vista, however, has been to improve the state of security in the Windows operating system. One common criticism of Windows XP and its predecessors has been their commonly exploited security ***vulnerabilities*** and overall ***susceptibility*** to **malware**, viruses and *buffer overflows. In light of this, Microsoft chairman Bill Gates announced in early 2002 a company-wide "Trustworthy Computing initiative" which aims to **incorporate** security work into every aspect of software development at the company*. Microsoft stated that it prioritized improving the security of Windows XP and Windows Server 2003 above finishing Windows Vista, thus delaying its completion.

While these new features and security improvements have ***garnered*** positive reviews, Vista has also been the target of much criticism and negative ***press***. Criticism of Windows Vista has targeted high system requirements, its more restrictive licensing terms, the inclusion of a number of new digital rights management technologies aimed at restricting the copying of protected digital media, lack of ***compatibility*** with certain pre-Vista hardware and software, and the number of ***authorization prompts*** for User Account Control. As a result of these and other issues, Windows Vista has seen adoption and satisfaction rates lower than Windows XP.

⊙Development

Microsoft began work on Windows Vista, known at the time by its codename Longhorn in May 2001, five months before the release of Windows XP. It was originally expected to ship sometime late in 2003 as a minor step between Windows XP and Blackcomb, which was planned to be the company's next major operating system release. Gradually, "Longhorn" ***assimilated*** many of the important new features and technologies slated for Blackcomb, resulting in the release date being pushed back several times. Many of Microsoft's developers were also re-***tasked*** to build updates to Windows XP and Windows Server 2003 to strengthen security. Faced with ongoing delays and concerns about feature ***creep***, Microsoft announced on August 27, 2004 that it had revised its plans. The original Longhorn, based on the Windows XP source code, was ***scrapped***, and Longhorn's development started ***anew***, building on the Windows Server 2003 Service Pack 1 ***codebase***, and re-incorporating only the features that would be intended for an actual operating system release. Some previously announced features such as WinFS were dropped or postponed, and a new software development methodology called the

Security Development Lifecycle was incorporated in an effort to address concerns with the security of the Windows codebase.

After Longhorn was named Windows Vista in July 2005, an unprecedented beta-test program was started, involving hundreds of thousands of volunteers and companies. In September of that year, Microsoft started releasing regular Community Technology Previews (**CTP**) to beta testers. The first of these was distributed at the 2005 Microsoft Professional Developers Conference, and was subsequently released to beta testers and Microsoft Developer Network subscribers. The *builds* that followed incorporated most of the planned features for the final product, as well as a number of changes to the user interface, based largely on feedback from beta testers. Windows Vista was *deemed* feature-complete with the release of the "February CTP", released on February 22, 2006, and much of the *remainder* of work between that build and the final release of the product focused on stability, performance, application and *driver* compatibility, and documentation. Beta 2, released in late May, was the first build to be made available to the general public through Microsoft's Customer Preview Program. It was downloaded by over five million people. Two release candidates followed in September and October, both of which were made available to a large number of users.

While Microsoft had originally hoped to have the consumer versions of the operating system available worldwide in time for Christmas 2006, it was announced in March 2006 that the release date would be pushed back to January 2007, in order to give the company – and the hardware and software companies which Microsoft depends on for providing device drivers – additional time to prepare. Through much of 2006, analysts and *bloggers* had speculated that Windows Vista would be delayed further, owing to *anti-trust* concerns raised by the European Commission and South Korea, and due to a *perceived* lack of progress with the beta releases. However, with the November 8, 2006 announcement of the completion of Windows Vista, Microsoft's lengthiest operating system development project came to an end.

Core Technologies

Windows Vista is intended to be a technology-based release, to provide a base to include advanced technologies, many of which are related to how the system functions and thus not readily visible to the user. An example is the complete restructuring of the architecture of the audio, print, display, and networking subsystems; although the results of this work are visible to software developers, end-users will only see what appear to be evolutionary changes in the user interface.

Vista includes technologies such as ReadyBoost and ReadyDrive which employ fast flash memory (located on USB drives and hybrid hard disk drives) to improve system performance by caching commonly used programs and data. This manifests itself in improved battery life on notebook computers as well, since a hybrid drive can be **spun** down when not in use. Another new technology called SuperFetch utilizes machine learning techniques to analyze usage patterns to allow Windows Vista to make intelligent decisions about what content should be present in system memory at any given time. It uses almost all the extra RAM as disk cache. *In conjunction with SuperFetch, an automatic built-in Windows Disk* **Defragmenter** *makes sure that those applications are strategically positioned on the hard disk where they can be loaded into memory very quickly with the least amount of physical movement of the hard disk's read-write heads.*

As part of the redesign of the networking architecture, IPv6 has been fully incorporated into the operating system and a number of performance improvements have been introduced, such as **TCP** window *scaling*. Earlier versions of Windows typically needed third-party wireless networking software to work properly, but this is not the case with Vista, which includes more comprehensive wireless networking support.

For graphics, Vista introduces a new Windows Display Driver Model and a major revision to Direct3D. The new driver model facilitates the new Desktop Window Manager, which provides the **tearing**-free desktop and special effects that are the **cornerstones** of Windows Aero. Direct3D 10, developed in conjunction with major display driver manufacturers, is a new architecture with more advanced **shader** support, and allows the graphics processing unit to render more complex scenes without assistance from the CPU. It features improved load balancing between CPU and GPU and also optimizes data transfer between them.

At the core of the operating system, many improvements have been made to the memory **manager**, process **scheduler** and I/O scheduler. *The* **Heap** *Manager implements additional features such as integrity checking in order to improve* **robustness** *and defend against buffer overflow security* **exploits**, *although this comes at the price of breaking backward compatibility with some* **legacy** *applications*. A Kernel Transaction Manager has been implemented that enables applications to work with the file system and **Registry** using **atomic** transaction operations.

Competition with Windows XP

According to a marketing manager working for HP Australia, Windows XP is still being chosen over Windows Vista for the majority of business computer sales. As all customers of **OEM** versions of Vista Business and Ultimate are **eligible** for a free downgrade to Windows XP Professional, these Windows XP licenses are sold as Vista Business licenses, thus increasing Vista's sales figures. Some computer manufacturers have chosen to ship Windows XP restore disks along with new computers with Vista Business and Ultimate editions pre-installed, as well as new computers with

XP instead of Vista.

 Words

vulnerability[ˌvʌlnərəˈbiləti] *n.*弱点	**blogger**[ˈblɔgə] *n.*写博客的人
susceptibility[səˌseptəˈbiliti] *n.*易感染性	**antitrust**[ˌæntiˈtrʌst] *adj.*反托拉斯的，反垄断的
incorporate[inˈkɔːpəreit] *v.*合并，加入	**perceive**[pəˈsiːv] *v.*察觉
garner[ˈgɑːnə] *v.*得到，收集	**spin**[spin] *v.*使快速旋转
press[pres] *n.*新闻报道，报刊	**defragmenter**[diːfrægməntə] *n.*磁盘碎片整理程序
compatibility[kəmˌpætiˈbiliti] *n.*兼容	**scale**[skeil] *v.*调节
authorization[ˌɔːθəraiˈzeiʃən] *n.*授权，认可	**tearing**[ˈtɛəriŋ] *adj.*令人难受的，猛烈的
prompt[prɔmpt] *n.*提示	**cornerstone**[ˈkɔːnəstəun] *n.*基础
assimilate[əˈsimileit] *v.*吸收	**shader**[ʃeidə] *n.*着色器
task[tɑːsk] *v.*分派任务	**manager**[ˈmænidʒə] *n.*管理器
creep[kriːp] *n.*蠕变	**scheduler**[ˈʃedjuːlə] *n.*调度程序，调节器
scrap[skræp] *v.*扔弃	**heap**[hiːp] *n.*堆
anew[əˈnjuː] *adv.*重新，再	**robustness**[rəˈbʌstnis] *n.*健壮性
codebase[ˈkəudˈbeis] *n.*代码库	**exploit**[iksˈplɔit] *n.*业绩，功绩
build[bild] *n.*构造	**legacy**[ˈlegəsi] *n.*遗产，遗留物
deem[diːm] *v.*认为	**Registry**[ˈredʒistri] *n.*注册表
remainder[riˈmeində] *n.*剩余物	**atomic**[əˈtɔmik] *adj.*原子的
driver[ˈdraivə] *n.*驱动器，驱动程序	**eligible**[ˈelidʒəbl] *adj.*符合条件的，合格的

 Phrases

peer-to-peer	对等，对等网络	**in conjunction with**	与…协力
buffer overflow	缓冲区溢出		

 Abbreviations

API	Application Programming Interface	应用编程接口
malware	Malicious Software	恶意软件
CTP	Community Technology Previews	社区技术预览版
TCP	Transfer Control Protocol	传输控制协议
OEM	Original Equipment Manufacturer	原始设备制造商

 Complex Sentences

1. **Original:** In light of this, Microsoft chairman Bill Gates announced in early 2002 a company-wide "Trustworthy Computing initiative" which aims to incorporate security work into every

aspect of software development at the company.

Translation: 鉴于这种情况，微软公司董事长Bill Gates在2002年初宣布了一个公司范围内的"可靠计算新方案"，旨在将安全工作纳入到公司软件开发的每个方面。

Exercises

I. Read the following statements carefully, and decide whether they are true (T) or false (F) according to the text.

____1. Windows Vista was begun to work after the release of Windows XP.

____2. Windows Vista can provide more comprehensive wireless networking support than its predecessors.

____3. Many of advanced technologies in Windows Vista are useful only for the professional developers.

____4. The consumer versions of Windows Vista were available worldwide for Christmas 2006.

____5. Windows Vista has exceeded its predecessors as the majority of business computer sales.

II. Choose the best answer to each of the following questions.

1. Which of the following is NOT the cause of delaying the completion of Windows Vista?

(A) Windows Vista assimilated many of the important new features and technologies slated for Blackcomb.

(B) Microsoft gave additional time for preparing to the companies which provide device drivers for it.

(C) Microsoft gave a priority to security improvement of Windows XP and Windows Server 2003

(D) Windows Vista had a longtime beta-test program to improve and complete features.

2. Which of the following employs fast flash memory to improve system performance in Windows Vista?

(A) ReadyBoost and ReadyDrive   (B) Windows Disk Defragmenter

(C) SuperFetch   (D) Windows Display Driver Model

3. Which of the following is NOT included in the new feature set in Windows Vista?

(A) Windows Aero   (B) WinFS   (C) Direct3D 10   (D) .NET Framework 3.0

III. Translating.

1. In conjunction with SuperFetch, an automatic built-in Windows Disk Defragmenter makes sure that those applications are strategically positioned on the hard disk where they can be loaded into memory very quickly with the least amount of physical movement of the hard disk's read-write heads.

2. The Heap Manager implements additional features such as integrity checking in order to improve robustness and defend against buffer overflow security exploits, although this comes at the price of breaking backward compatibility with some legacy applications.

⨠ Section B   Linux

Linux is a Unix-like computer operating system family which uses the Linux kernel. Linux is one of the most *prominent* examples of free software and open source development; typically all the underlying source code can be freely modified, used, and redistributed by anyone.

Predominantly known for its use in servers, it is installed on a wide variety of computer hardware, ranging from embedded devices and mobile phones to supercomputers.

While the user base of Linux is small, many users are expert and active programmers; the effects of Linux extend well beyond the user base. Software developed on Linux can run with little or no alteration on other Unix machines, e.g. Mac OS X, many Websites are hosted on **LAMP** stacks, and many technologies originating in the Linux domain are eventually *ported* to Windows.

Several corporations are deeply involved, with a *vested* interest in Linux, and many more give financial and other support such as Red Hat, Novell, Oracle Corporation, Hewlett-Packard, IBM, Sun Microsystems, Nokia, and Dell.

The name "Linux" comes from the Linux kernel, originally written in 1991 by Linus Torvalds. The system's *utilities* and libraries usually come from the **GNU** operating system, announced in 1983 by Richard Stallman. The GNU contribution is the basis for the alternative name GNU/Linux.

⊖ History

The Unix operating system was conceived and implemented in the 1960s and first released in 1970. Its wide availability and portability meant that it was widely adopted, copied and modified by academic institutions and businesses, with its design being influential on authors of other systems.

The GNU Project, started in 1984 by Richard Stallman, had the goal of creating a "complete Unix-compatible software system" made entirely of free software. The next year Stallman created the Free Software Foundation and wrote the GNU General Public License (**GNU GPL**) in 1989. By the early 1990s, many of the programs required in an operating system (such as libraries, *compilers*, text editors, a Unix shell, and a windowing system) were completed, although low-level elements such as device drivers, *daemons*, and the kernel were *stalled* and incomplete. Linus Torvalds has said that if the GNU kernel had been available at the time (1991), he would not have decided to write his own.

⊖ MINIX

MINIX, a Unix-like system intended for academic use, was released by Andrew S. Tanenbaum in 1987. While *source code* for the system was available, modification and redistribution was restricted (that is not the case today). In addition, MINIX's 16-bit design was not well adapted to the 32-bit design of the increasingly cheap and popular Intel 386 architecture for personal computers.

In 1991, Torvalds began to work on a non-commercial replacement for MINIX while he was attending the University of Helsinki., which would eventually become the Linux kernel.

In 1992, Tanenbaum posted an article on Usenet claiming Linux was obsolete. *In the article, he criticized the operating system as being **monolithic** in design and being tied closely to the x86 architecture and thus not portable, as he described "a fundamental error."* Tanenbaum suggested that those who wanted a modern operating system should look into one based on the microkernel model. The posting **elicited** the response of Torvalds, which resulted in a well known debate over the microkernel and monolithic kernel designs.

Linux was dependent on the MINIX user space at first. With code from the GNU system freely available, it was advantageous if this could be used with the **fledgling** OS. Code licensed under the GNU GPL can be used in other projects, so long as they also are released under the same or a compatible license. In order to make the Linux kernel compatible with the components from the GNU Project, Torvalds initiated a switch from his original license (which prohibited commercial redistribution) to the GNU GPL. Linux and GNU developers worked to integrate GNU components with Linux to make a fully functional and free operating system.

Commercial and Popular *Uptake*

*Today Linux is used in numerous domains, from embedded systems to supercomputers, and has **secured** a place in server installations with the popular LAMP application stack.* Torvalds continues to direct the development of the kernel. Stallman heads the Free Software Foundation, which in turn supports the GNU components. Finally, individuals and corporations develop third-party non-GNU components. These third-party components comprise a vast body of work and may include both kernel modules and user applications and libraries. Linux vendors and communities combine and distribute the kernel, GNU components, and non-GNU components, with additional package management software in the form of Linux *distributions*.

Development

A summarized history of Unix-like operating systems shows Linux's origins. Note that despite similar architectural designs and concepts being shared as part of the **POSIX** standard, Linux does not share any non-free source code with the original Unix or Minix.

The primary difference between Linux and many other popular contemporary operating systems is that the Linux kernel and other components are free and open source software. Linux is not the only such operating system, although it is the best-known and most widely used. Some free and open source software licenses are based on the principle of *copyleft*, a kind of *reciprocity*: any work derived from a copyleft piece of software must also be copyleft itself. The most common free software license, the GNU GPL, is a form of copyleft, and is used for the Linux kernel and many of the components from the GNU project.

As an operating system ***underdog*** competing with mainstream operating systems, Linux cannot rely on a ***monopoly*** advantage; in order for Linux to be convenient for users, Linux aims for interoperability with other operating systems and has established computing standards. Linux systems adhere to POSIX, **SUS**, **ISO** and **ANSI** standards where possible, although *to date* only one Linux distribution has been POSIX.1 certified, Linux-FT.

Free software projects, although developed in a collaborative fashion, are often produced independently of each other. *However, given that the software licenses **explicitly** permit redistribution, this provides a basis for larger scale projects that collect the software produced by stand-alone projects and make it available all at once in the form of a Linux distribution.*

A Linux distribution, commonly called a "distro", is a project that manages a remote collection of Linux-based software, and facilitates installation of a Linux operating system. Distributions are maintained by individuals, *loose-knit* teams, volunteer organizations, and commercial entities. They include system software and application software in the form of packages, and distribution—specific software for initial system installation and configuration as well as later package upgrades and installs. A distribution is responsible for the default configuration of installed Linux systems, system security, and more generally integration of the different software packages into a coherent whole.

Market Share and Uptake

Many quantitative studies of open source software focus on topics including market share and reliability, with numerous studies specifically on examining Linux. The Linux market is growing rapidly, and the revenue of servers, desktops, and packaged software running Linux is expected to exceed $35.7 billion by 2008.

IDC's report for Q1 2007 says that Linux now holds 12.7% of the overall server market. This estimate was based on the number of Linux servers sold by various companies.

Estimates for the desktop market share of Linux range from less than one percent to almost two percent. In comparison, Microsoft operating systems hold more than 90%.

The frictional cost of switching operating systems and lack of support for certain hardware and application programs designed for Microsoft Windows have been two factors that have inhibited adoption. *Proponents* and analysts attribute the relative success of Linux to its security, reliability, low cost, and freedom from vendor *lock-in*.

Also most recently Google has begun to fund Wine, which acts as a compatibility layer, allowing users to run some Windows programs under Linux.

The XO laptop project of One Laptop Per Child is creating a new and potentially much larger Linux community, planned to reach several hundred million schoolchildren and their families and communities in developing countries. Six countries have ordered a million or more units each for delivery in 2007 to distribute to schoolchildren at no charge. Google, Red Hat, and eBay are major supporters of the project. While the XO will also have a Windows option, it will be primarily deployed using **RHEL**.

Copyright and Naming

The Linux kernel and most GNU software are licensed under the GNU General Public License (GPL). The GPL requires that anyone who distributes the Linux kernel must make the source code (and any modifications) available to the recipient under the same terms. In 1997, Linus Torvalds stated, "Making Linux GPL was definitely the best thing I ever did." Other key components of a Linux system may use other licenses; many libraries use the GNU Lesser General Public License (**LGPL**), a more permissive variant of the GPL, and the X Window System uses the MIT License.

Torvalds has publicly stated that he would not move the Linux kernel (currently licensed under GPL version 2) to version 3 of the GPL, released in mid-2007, specifically citing some *provisions* in the new license which prohibit the use of the software in digital rights management.

A 2001 study of Red Hat Linux 7.1 found that this distribution contained 30 million source lines of code. Using the Constructive Cost Model, the study estimated that this distribution required about eight thousand man-years of development time. According to the study, if all this software had been developed by conventional *proprietary* means, it would have cost about 1.08 billion dollars (year 2000 U.S. dollars) to develop in the United States.

Most of the code (71%) was written in the C programming language, but many other languages were used, including C++, *assembly language*, Perl, Python, Fortran, and various shell scripting languages. Slightly over half of all lines of code were licensed under the GPL. The Linux kernel itself was 2.4 million lines of code, or 8% of the total.

In a later study, the same analysis was performed for Debian GNU/Linux version 4.0. This distribution contained over 283 million source lines of code, and the study estimated that it would have cost 5.4 billion Euros to develop by conventional means.

In the United States, the name Linux is a trademark registered to Linus Torvalds. Initially, nobody registered it, but on 15 August 1994, William R. Della Croce, Jr. *filed* for the trademark Linux, and then demanded *royalties* from Linux distributors. In 1996, Torvalds and some affected organizations sued him to have the trademark assigned to Torvalds, and in 1997 the case was settled. The licensing of the trademark has since been handled by the Linux Mark Institute. Torvalds has stated that he only trademarked the name to prevent someone else from using it, but was *bound* in 2005 by United States trademark law to take active measures to *enforce* the trademark. As a result, the **LMI** sent out a number of letters to distribution vendors requesting that a fee be paid for the use of the name, and a number of companies have *complied*.

GNU/Linux

The Free Software Foundation views Linux distributions which use GNU software as GNU variants and they ask that such operating systems be referred to as GNU/Linux or a Linux-based GNU system. However, the media and population *at large* refers to this family of operating systems simply as Linux. Some distributions use GNU/Linux (particularly notable is Debian GNU/Linux), but the term's use outside of the *enthusiast* community is limited. *The distinction between the Linux kernel and distributions based on it plus the GNU system is a source of confusion to many newcomers, and the naming remains controversial, as many large Linux distributions (e.g. Ubuntu and SuSE Linux) are simply using the Linux name, rather than GNU/Linux.*

 Words

prominent['prɔminənt] *adj.*著名的，突出的
predominantly[pri'dɔminəntli] *adj.*最显著的，最有影响的
port[pɔ:t] *v.*移植
vested['vestid] *adj.*既定的，确定的
utility[ju:'tiliti] *n.*实用程序，应用程序
compiler[kəm'pailə] *n.*编译器
daemon['di:mən] *n.*Unix和其他多任务处理操作系统中一种在后台运行的计算机程序
stall[stɔ:l] *v.*停止，迟延
monolithic[,mɔnə'liθik] *adj.*庞大的，完整的
elicit[i'lisit] *v.*得出
fledgling['fledʒliŋ] *adj.*年轻的或无经验的
uptake['ʌpteik] *n.*理解，举起
secure[si'kjuə] *v.*获得

distribution[,distri'bju:∫ən] *n.*销售版本，销售形式
copyleft['kɔpileft] *v.*非盈利版权
reciprocity[,risi'prɔsiti] *n.*互惠
underdog['ʌndədɔg] *n.*失败者，受压迫者
monopoly[mə'nɔpəli] *n.*垄断，专利权
explicitly[iks'plisitli] *adv.*清楚地，明晰地
proponent[prə'pəunənt] *n.*建议者，支持者
provision[prə'viʒən] *n.*规定，条款
proprietary[prə'praiətəri] *n.*专有的，专卖的
file[fail] *v.*提出（申请等），呈请把…备案
royalties['rɔiəlti] *n.*版税
bind[baind] *v.*使受法律（或合同、道义等）约束
enforce[in'fɔ:s] *v.*实施，执行
comply[kəm'plai] *v.*遵从，遵照
enthusiast[in'θju:ziæst] *n.*热心家，狂热者

Phrases

source code	源代码	lock-in	锁定
to date	到目前为止	assembly language	汇编语言
given that...	假定，已知	at large	随便地，笼统地
loose-knit	可拆开的		

Abbreviations

LAMP	Linux, Apache, MySQL and PHP	一组常用来搭建动态网站或者服务器的开源软件
GNU	GNU's Not Unix	一个完全由自由软件组成的计算机操作系统
GNU GPL	GNU General Public License	GNU通用公共许可证
POSIX	Portable Operating System Interface of Unix	Unix可移植操作系统接口
SUS	Single UNIX Specification	单一UNIX规范
ISO	International Organization for Standardization	国际标准化组织
ANSI	American National Standards Institute	美国国家标准协会
RHEL	Red Hat Enterprise Linux	Red Hat企业版Linux
LGPL	Lesser General Public License	较宽松通用公共许可证
LMI	Linux Mark Institute	Linux商标协会

Complex Sentences

1. **Original:** In the article, Tanenbaum criticized the operating system as being monolithic in design and being tied closely to the x86 architecture and thus not portable, as he described "a fundamental error."
   Translation: 在这篇文章中，Tanenbaum批评操作系统在设计上完整而庞大，并且紧密依赖x86体系结构而不可移植，正像他所描述的，这是"一个根本性的错误"。

2. **Original:** Today Linux is used in numerous domains, from embedded systems to supercomputers, and has secured a place in server installations with the popular LAMP application stack.
   Translation: 现在Linux应用于众多领域，从嵌入式系统到超级计算机，并且已经在用流行的LAMP应用程序栈进行服务器安装方面获得了一席之地。

Exercises

I. **Read the following statements carefully, and decide whether they are true (T) or false (F) according to the text.**

____1. The Linux kernel derived from Torvalds's work on a non-commercial replacement for MINIX.

____2. The alternative name GNU/Linux comes from Linux using GNU kernel.

____3. Linux distributions include system software, application software and distribution-specific software.

___4. Red Hat Linux 7.1 cost about 1.08 billion dollars and about eight thousand man-years of development time to develop.

___5. The name of free software is prohibited to be registered as a trademark or require royalties from Linux distributors.

II. Choose the best answer to each of the following questions.

1. Which of the following was released first among others?

(A) Linux   (B) Unix   (C) MINIX   (D) GNU

2. Which of the following is WRONG about Linux and Windows?

(A) Linux lacks support for certain hardware and application programs designed for Microsoft Windows.

(B) Linux is successfuldue to its security, reliability, low cost, and freedom from vendor lock-in.

(C) As a compatibility layer, Wine allows users to run some Linux programs under Windows.

(D) Many technologies originating in the Linux domain are eventually ported to Windows.

3. Which of the following is RIGHT about free and open source software?

(A) The source code of the software can be freely modified, used, and redistributed by anyone.

(B) The kernel and key components of the software must be licensed under the same License.

(C) Any work on the kernel and other components are based on the principle of non-copyright.

(D) Every free software projects are developed and produced in a centralized fashion.

III. Translating.

1. However, given that the software licenses explicitly permit redistribution, this provides a basis for larger scale projects that collect the software produced by stand-alone projects and make it available all at once in the form of a Linux distribution.

2. The distinction between the Linux kernel and distributions based on it plus the GNU system is a source of confusion to many newcomers, and the naming remains controversial, as many large Linux distributions (e.g. Ubuntu and SuSE Linux) are simply using the Linux name, rather than GNU/Linux.

Part 2   Simulated Writing: Meeting Minutes

Introduction

Minutes are written as an accurate record of a group's meetings. They record the decisions of the meeting and the actions agreed and, importantly, provide a review document for use at the next meeting so that progress can be measured – this makes them a useful disciplining technique as individuals' performance and non-performance of agreed actions is given high visibility. Minutes can also inform people who were not at the meeting about what took place.

⊝ Contents

Before each meeting an agenda should be drawn up, detailing the matters to be discussed at the meeting. A set of minutes should normally include the following information:

- Time, date and place of meeting
- List of people attending, and list of absent members of the group
- Approval of the previous meeting's minutes, and any matters arising from those minutes
- For each item in the agenda, a record of the principal points discussed and decisions taken
- List of actions agreed upon
- For each item in the actions, a list of person responsible and deadline
- Time, date and place of next meeting
- Name of person taking the minutes

⊝ Writing Tips

There are some tips for you to write an effective meeting minutes:

- Distribute (by email) the agenda before the meeting, so that members of the group have a chance to prepare for the meeting.
- Include an item "AOB" (Any Other Business) at the end of the agenda as a place to include last-minute items.
- Keep the minutes short and to the point. Don't waffle. If you want to record every word said, you might consider a tape recording to supplement the minutes.
- Where a member of the group is asked to perform a set task, record an "Action" point; this makes it easy to read through the minutes at the next meeting and " tick off " the action points.
- Either write the minutes as the meeting happens (if the minutes secretary is a fast typist), or immediately after the meeting. The sooner they are done, the more accurate they are.
- Write minutes in the past tense. You are writing about discussions that have already happened. When you are typing up the minutes from your notes, you are recording a past event.

⊝ Format

Name of Organization
Month Day, Year
Time and location
Present: Name members in attendance
Absent: Name absent members.

Proceedings:

- Meeting called to order at (time) by (person, usually chair)
- Minutes from (prior meeting date) amended and approved
- Highlights of information presented
- Any action taken
- Meeting adjourned at (time)

Future Business:

Here is a place to remind people of:

- conversations that were tabled until next time
- possible agendas for upcoming meetings
- assignments that members have taken on

Minutes submitted by (name)

Sample

FLY 3D Game
Minutes of the 6th Group Meeting

Date: October 4, 2008
Time: 9:30 a.m. - 10:00 a.m.
Duration: 0.5 hour
Venue: Prof. Smith's office
Present: Prof. Smith,
　　　　　Chau Chun Ting (Charles),
　　　　　Chang Kin Fung (Tony)
Minutes recorder: Chang Kin Fung (Tony)
Absent: Lam Sheung Yan (Michael)
　　　　Au Kwok Wang (Chris)

1. Approval of Minutes of the Last Meeting
- Minutes of the last meeting were approved as an accurate record.

2. Discussion of Project Development
- Charles raised the question about the camera: if the camera is above the character at some particular angle, then we are unable to see very far to the front and may not see the enemies.
- Professor Smith said that there was no so-called "good" view angle. If the camera view is that of the character, then we can see the virtual world, but we are unable to see the

character. The player may lose his orientation since he has no sense about where the character is.

- Tony asked whether it was a problem for the character to turn around because the scene would change very quickly and thus make the player feel uncomfortable.
- Prof. Smith said that limiting the speed of turnaround could solve the problem. He said that the main point was to make the game interesting and exciting. The view angle was not that important.
- Professor Smith suggested that we should design a map to display the location of the character so that the players can be aware of the progress and the place of the character.

3. Meeting Arrangements

- Professor Smith asked each group to give a simple demo of their FLY in the next group meeting.
- For the demo, each group should be able to implement a 3D environment with the character's movements, for instance, forward and backward movements.
- The purpose of the demo is to make sure each group has some basic ideas of how to implement 3D objects and control them.

4. Adjournment of Meeting

The meeting was adjourned at 10:00 a.m.

5. Next Meeting

Date: October 12, 2008
Time: 2:00 p.m. - 3:00 p.m.
Place: FLY Lab

6. Actions Agreed Upon

Action list, dated 4 October 2008					
Item no.	Action	By	Deadline	Status	
1	Give a simple demo of a 3D sky environment with the character's movements	Charles	12 Oct. 2008		
2	Give a simple demo of a 3D water environment with the character's movements	Tony	12 Oct. 2008		
3	Give a simple demo of a 3D land environment with the character's movements	Patrick	12 Oct. 2008		
4	Give a simple demo of a character	Chris	12 Oct. 2008		

Written by Tony on October 5, 2008

Part 3 Listening and Speaking

Dialogue: Choosing a Linux Distribution and Free Trials for Free Software

(On the first lesson of Introduction to Linux, Mr. White called for every student to choose and install a Linux distribution on their computers as an assignment. Henry asked for his friends, Mark and Sophie about it after class.)

Henry	With hundreds of Linux distributions available[1], choosing one seems hard.
Sophie	Don't worry, Henry. If you're not sure which distribution is right for you yet, you can do some comparison between them first.
Mark	Yes. Above all, there are two main sorts of distributions, special purpose and general purpose desktop distribution. Specialized distributions solve particular problems for software development, security, or user support. Desktop distributions are designed for regular users. [2]
Henry	Other than this, what are the differences between commercial and noncommercial distribution[3]?
Sophie	Linspire, SUSE, Mandrakesoft, and other commercial desktop vendors offer installation support, printed manuals, and support Websites. Vendors of noncommercial distributions may not.
Mark	Commercial desktop distributions are almost all priced at less than $100, many less than $50, and all include the equivalent of hundreds of dollars' worth of commercial applications for word processing, productivity, financial, email, Web, and other Internet and networking software. [4]
Sophie	In addition, do you need an easier or professional distribution? Easier means fewer choices. Professional distributions from SUSE or Mandrakesoft will satisfy *power* users with more software choices. [5]
Henry	As a beginner, I think an easier distribution is not difficult to learn and more fit for me.
Sophie	That's right. Entry-level, all-in-one distributions, such as those from Lycoris and Linspire, can be extraordinarily easy to use.

Mark	*As yet*, if you're not ready to *fiddle around* with a PC that works fine but you still want to give Linux a try, there sare many Linux vendors that offer distributions on **bootable** CDs or **ISOs**.
Henry	Really? How can I get them?
Mark	Many Linux installations are available as ISO downloads, so you only need to download a single file instead of downloading many smaller files and then combining them all with software. Just pop the CD/ISO into any CD-bootable system. You have a fully functioning Linux PC. Most distributions are freely downloadable in some form, so you can try as many as you like. [6]
Sophie	Most consumer-oriented Linux vendors offer these tools for such purposes as testing, or turning any PC into a Linux PC in about five minutes. Relatively slow CD drives limit performance, as well as the selection of applications, but there are enough applications to browse the Web, write documents, check emails, and do most of the things you need from a PC. You can get free trials for free software, and decide later which software you want to install.
Henry	Ok, I've got it! Thank you very much for your valuable advice.
Mark & Sophie	Good luck!

Exercises

Work in pairs, and make up a similar conversation by replacing the scenarios with other material in the below.

Choosing an Edition from the Windows Family

[1] A big Windows family tree and many editions of each line

[2] Microsoft has taken two parallel routes in its operating systems. One route has been for the home user and the other has been for the professional IT user. The dual routes have generally led to home versions having greater multimedia support and less functionality in networking and security, and professional versions having inferior multimedia support and better networking and security.

[3] Windows XP and Windows Server 2003

[4] Windows XP is a line of operating systems for use on personal computers, including home and business desktops, notebook computers, and media centers.
Windows Server is the platform for building an infrastructure of connected applications,

networks, and Web services, from the workgroup to the data center.

[5] The two major editions are Windows XP Home Edition, designed for home users, and Windows XP Professional, designed for business and power-users. XP Professional contains advanced features that the average home user would not use. A third edition, called Windows XP Media Center Edition incorporates new digital media, broadcast television and Media Center Extender capabilities.

[6] These releases were made available for download from Microsoft's Web site or at retail outlets that sell computer software, and were pre-installed on computers sold by major computer manufacturers.

 Words

power['pauə] *adj.*	专业的，有影响力的	bootable['bu:təbl] *adj.*	可引导的

 Phrases

call for	要求，提倡	as yet	到目前为止
ask for	请求，寻找	fiddle around	不经意地干活，闲荡

 Abbreviations

ISO	ISOlation	一种镜像文件

Listening Comprehension: Open Source Software

Listen to the passage and the following 3 questions based on it. After you hear a question, there will be a break of 10 seconds. During the break, you will decide which one is the best answer among the four choices marked (A), (B), (C) and (D).

Questions

1. Which of the following items is NOT the characteristic of Open Source Software according to this article?

2. How many noted Open Source projects in the world today are mentioned in this article?

3. What is the author's attitude towards Open Source Software in this article?

Choices

1. (A) Reproducible   (B) Improvable   (C) Collaborated   (D) Unsellable

2. (A) Three   (B) Four   (C) Five   (D) Six

3. (A) Positive   (B) Negative   (C) Critical   (D) Caustic

 Words

> **duplicable**['dju:plikəbl] *adj.*可复制的，可再发生的
> **derivative**[di'rivətiv] *n.*衍生物，派生的事物
> **irreversible**[ˌiri'və:səbl] *adj.*不可改变的，不可逆的

>> Dictation: Apple Mac OS

This passage will be played THREE times. Listen carefully, and fill in the blanks with the words you have heard.

Mac OS is a line of graphical user interface-based operating systems developed by Apple Inc., who ***deliberately downplayed*** the _____1_____ of the operating system in the early years of the ***Macintosh*** to help make the machine _____2_____ more user-friendly and to distance it from other operating systems such as MS-DOS, which were ***portrayed*** as ***arcane*** and technically _____3_____. The Mac OS is characterized by its total lack of a _____4_____ ; it is a _____5_____ graphical operating system.

The Mac OS can be divided into two _____6_____ of operating systems:

"Classic" Mac OS was shipped with the first Macintosh in 1984 and its ***descendants***, ***culminating*** with Mac OS 9. Much of this early system software was held in ROM on the _____7_____. The initial _____8_____ of this was to avoid using up the limited _____9_____ of floppy disks on system support, given that the early Macs had no hard disk. This _____10_____ also allowed for a completely graphical OS _____11_____ at the lowest level without the need for a text-only ***console*** or command-line mode. A ***fatal*** software error, or even a low-level hardware error discovered during system _____12_____, was _____13_____ to the user graphically using some combination of _____14_____, alert box windows, buttons, a mouse _____15_____, and the distinctive Chicago bitmap font. Mac OS depended on this _____16_____system software in ROM, a fact that later helped to ensure that only Apple computers or licensed clones could run Mac OS.

The newer **Mac OS X** (where the X is 10 written as a Roman numeral) is the successor to the original Mac OS. Bringing Unix-style memory _____17_____ and ***pre-emptive*** multitasking to the Mac platform, Mac OS X is based on the Mach kernel and the **BSD** _____18_____ of UNIX. Its new memory management system allowed more programs to run at once and virtually eliminated the _____19_____ of one program crashing another. The latest of Mac OS X is pre-loaded on all _____20_____ shipping Macintosh computers.

Words

deliberately[di'libərətli] *adv.*故意地

downplay ['daunplei] *v.*不予重视

Macintosh ['mækin,tɔʃ] *n.* Apple公司于
　　1984年推出的一种系列微机，简称Mac

portray [pɔː'trei] *v.*描绘

arcane [ɑː'kein] *adj.*晦涩难解的，神秘的

descendant [di'send(ə)nt] *n.*子孙，后代

culminate ['kʌlmineit] *v.*达到顶点，告终

console ['kɔnsəu] *n.*控制台，操纵台

fatal ['feitl] *adj.*致命的，毁灭性的

pre-emptive [priː'emptiv] *adj.*抢先的，有先买权的

 Abbreviations

| BSD | Berkeley Software Distribution | 伯克利软件套件 |

Computer Programming

Part 1   **Reading and Translating**

Section A   Object-Oriented Programming

Object-oriented programming (**OOP**) is a programming *paradigm* that uses "objects" and their interactions to design applications and computer programs. Programming techniques may include features such as encapsulation, *modularity*, *polymorphism*, and *inheritance*. It was not commonly used in mainstream software application development until the early 1990s. Many modern programming languages now support OOP.

Introduction

Object-oriented programming can trace its roots to the 1960's. As hardware and software became increasingly complex, quality was often *compromised*. Researchers studied ways in which software quality could be maintained. *Object-oriented programming was deployed in part as an attempt to address this problem by strongly emphasizing* **discrete** *units of programming logic and* *re-usability in software. Computer programming methodology focuses on data rather than processes, with programs composed of self-sufficient modules (objects) containing all the information needed to manipulate a data structure.*

Object-oriented programming may be seen as a collection of cooperating objects, as opposed to a traditional view in which a program may be seen as a group of tasks to compute ("*subroutines*"). In OOP, each object is capable of receiving messages, processing data, and sending messages to other objects.

Fundamental Concepts

A survey by Deborah J. Armstrong of nearly 40 years of computing literature identified a number of "*quarks*", or fundamental concepts, found in the strong majority of definitions of OOP. They are the following:

1. Class

Class defines the abstract characteristics of a thing (object), including the thing's characteristics (its attributes, fields or properties) and the thing's behaviors (the things it can do, or methods, operations or features). For example, the class Dog would consist of traits shared by all dogs, such as *breed* and fur color (characteristics), and the ability to bark and sit (behaviors). Classes provide modularity and structure in an object-oriented computer program. A class should typically be recognizable to a non-programmer familiar with the problem domain, meaning that the characteristics of the class should make sense in context. Also, the code for a class should be relatively self-contained (generally using encapsulation). Collectively, the properties and methods defined by a class are called members.

2. Object

Object is a pattern (exemplar) of a class. The class of Dog defines all possible dogs by listing the characteristics and behaviors they can have; the object Lassie is one particular dog, with particular versions of the characteristics. A Dog has fur; Lassie has brown-and-white fur.

3. Instance

One can have an *instance* of a class or a particular object. The instance is the actual object created at runtime. In programmer *jargon*, the Lassie object is an instance of the Dog class. The set of values of the attributes of a particular object is called its state. The object consists of state and behavior that is defined in the object's class.

4. Method

Method is an object's ability. Lassie, being a Dog, has the ability to bark. So bark() is one of Lassie's methods. She may have other methods as well, for example sit(), eat(), walk() or save_timmy(). Within the program, using a method usually affects only one particular object; all Dogs can bark, but you need only one particular dog to bark.

5. Inheritance

"Subclasses" are more specialized versions of a class, which inherit attributes and behaviors from their parent classes, and can introduce their own. For example, the class Dog might have sub-classes called *Collie*, *Chihuahua*, and Golden*Retriever*. In this case, Lassie would be an instance of the Collie subclass. Suppose the Dog class defines a method called bark() and a property called furColor. Each of its sub-classes (Collie, Chihuahua, and GoldenRetriever) will inherit these members, meaning that the programmer only needs to write the code for them once.

Each subclass can alter its inherited traits. For example, the Collie class might specify that the default furColor for a collie is brown-and-white. The Chihuahua subclass might specify that the bark() method produces a high pitch by default. Subclasses can also add new members. The Chihuahua subclass could add a method called tremble(). So an individual Chihuahua instance would use a high-pitched bark() from the Chihuahua subclass, which in turn inherited the usual bark() from Dog. The Chihuahua object would also have the tremble() method, but Lassie would not, because she is a Collie, not a Chihuahua. In fact, inheritance is an "is-a" relationship: Lassie is a Collie. A Collie is a Dog. Thus, Lassie inherits the methods of both Collies and Dogs.

Multiple inheritance is inheritance from more than one ancestor class, neither of these ancestors being an ancestor of the other. For example, independent classes could define Dogs and Cats, and a *Chimera* object could be created from these two which inherits all the (multiple) behavior of cats and dogs. This is not always supported, as it can be hard both to implement and to use well.

6. Abstraction

Abstraction is simplifying complex reality by modeling classes appropriate to the problem, and working at the most appropriate level of inheritance for a given aspect of the problem. For example, Lassie may be treated as a Dog much of the time, a Collie when necessary to access Collie-specific attributes or behaviors, and as an Animal (perhaps the parent class of Dog) when counting Timmy's pets.

7. Encapsulation

Encapsulation conceals the functional details of a class from objects that send messages to it. For example, the Dog class has a bark() method. The code for the bark() method defines exactly how a bark happens (e.g., by *inhale*() and then *exhale*(), at a particular pitch and volume). Timmy, Lassie's friend, however, does not need to know exactly how she barks. Encapsulation is achieved by specifying which classes may use the members of an object. The result is that each object exposes to any class a certain interface — those members accessible to that class. The reason for encapsulation is to prevent clients of an interface from depending on those parts of the implementation that are likely to change in future, thereby allowing those changes to be made more easily, that is, without changes to clients. For example, an interface can ensure that *puppies* can only be added to an object of the class Dog by code in that class. Members are often specified as public, protected or private, determining whether they are available to all classes, sub-classes or only the defining class. *Some languages go further: Java uses the default access modifier to restrict access also to classes in the same package, C# and VB.NET reserve some members to classes in the same assembly using keywords internal (C#) or Friend (VB.NET), and Eiffel and C++ allow one to specify which classes may access any member.*

8. Polymorphism

Polymorphism in object-oriented programming is the ability of objects belonging to different data types to respond to method calls with the same name, each one according to an appropriate type-specific behavior. For example, if a Dog is commanded to speak(), this may elicit a bark().

9. Decoupling

Decoupling allows for the separation of object interactions from classes and inheritance into distinct layers of abstraction. A common use of decoupling is to polymorphically decouple the encapsulation, which is the practice of using reusable code to prevent discrete code modules from interacting with each other.

Not all of the above concepts can be supported in all object-oriented programming languages, and so object-oriented programming that uses classes is called sometimes class-based programming. In particular, prototype-based programming does not typically use classes. As a result, a significantly different yet *analogous* terminology is used to define the concepts of object and instance, although there are no objects in these languages.

History

The concept of objects and instances in computing had its first major breakthrough with the PDP-1 system at MIT which was probably the earliest example of capability-based architecture. Another early example was Sketchpad made by Ivan Sutherland in 1963; however, this was an application and not a programming paradigm. Objects as programming entities were introduced in the 1960s in Simula 67, a programming language designed for making *simulations*, created by Ole-Johan Dahl and Kristen Nygaard of the Norwegian Computing Center in Oslo.

(Reportedly, the story is that they were working on ship simulations, and were **confounded** by the combinatorial explosion of how the different attributes from different ships could affect one another. The idea occurred to group the different types of ships into different classes of objects, each class of objects being responsible for defining its own data and behavior.) Such an approach was a simple *extrapolation* of concepts earlier used in analog programming. On analog computers, such direct *mapping* from real-world phenomena/objects to analog phenomena/objects (and conversely) was (and is) called "simulation". Simula (Simulation Language) not only introduced the notion of classes, but also of instances of classes, which is probably the first explicit use of those notions. The ideas of Simula 67 influenced many later languages, especially Smalltalk and derivatives of Lisp and Pascal.

The Smalltalk language, which was developed at Xerox **PARC** in the 1970s, introduced the term object-oriented programming to represent the *pervasive* use of objects and messages as the basis for computation. Smalltalk creators were influenced by the ideas introduced in Simula 67, but Smalltalk was designed to be a fully dynamic system in which classes could be created and modified dynamically rather than statically as in Simula 67. Smalltalk and its OOP features were introduced to a wider audience by the August 1981 issue of Byte magazine.

Object-oriented programming developed as the dominant programming methodology during the mid-1990s, largely due to the influence of C++. Its dominance was further enhanced by the rising popularity of graphical user interfaces, for which object-oriented programming is well-suited. An example of a closely related dynamic GUI library and OOP language can be found in the Cocoa frameworks on Mac OS X, written in Objective C, an object-oriented, dynamic messaging extension to C based on Smalltalk. OOP toolkits also enhanced the popularity of event-driven programming (although this concept is not limited to OOP). Some feel that association with GUIs (real or perceived) was what *propelled* OOP into the programming mainstream.

At ETH Zürich, Niklaus Wirth and his colleagues had also been investigating such topics as data abstraction and modular programming. Modula-2 included both, and their succeeding design, Oberon, included a distinctive approach to object orientation, classes, and such. The approach is unlike Smalltalk, and very unlike C++.

Object-oriented features have been added to many existing languages during that time, including Ada, BASIC, Fortran, Pascal, and others. Adding these features to languages that were not initially designed for them often led to problems with compatibility and maintainability of code.

In the past decade Java has emerged in wide use partially because of its similarity to C and to C++, but perhaps more importantly because of its implementation using a virtual machine that is intended to run code unchanged on many different platforms. This last feature has made it very attractive to larger development shops with heterogeneous environments. Microsoft's .NET initiative has a similar objective and includes/supports several new languages, or variants of older ones with the important *caveat* that it is, of course, restricted to the Microsoft platform.

More recently, a number of object-oriented languages compatible with procedural methodology, such as Python and Ruby, have emerged. Besides Java, probably the most commercially important recent object-oriented languages are Visual Basic .NET (VB.NET) and C#, both designed for Microsoft's .NET platform. VB.NET and C# both support cross-language inheritance, allowing classes defined in one language to subclass classes defined in the other language.

Just as procedural programming led to refinements of techniques such as structured programming, modern object-oriented software design methods include refinements such as the use of design patterns, design by contract, and modeling languages (such as **UML**).

 Words

paradigm['pærədaim] *n*.范例	**retriever**[ri'tri:və] *n*.一种能把猎物找回来的猎犬
modularity[,mɔdju'læriti] *n*.模块性	**chimera**[kai'miərə] *n*.嫁接杂种
polymorphism[,pɔli'mɔ:fizəm] *n*.多态	**inhale**[in'heil] *v*.吸气
inheritance[in'heritəns] *n*.继承	**exhale**[eks'heil] *v*.呼气
compromise['kɔmprəmaiz] *v*.妥协,折中,危及…的安全	**puppy**['pʌpi] *n*.小狗,幼犬
	decouple[di'kʌpl] *v*.分离,消除…间相互影响
discrete[dis'kri:t] *adj*.分离的,离散的	**analogous**[ə'næləgəs] *adj*.类似的,相似的
subroutine[,sʌbru:'ti:n] *n*.子程序	**simulation**[,simju'leiʃən] *n*.仿真,模拟
quark[kwɑ:k] *n*.夸克(理论上一种比原子更小的基本粒子)	**confound**[kən'faund] *v*.使困惑,混乱,混淆
	extrapolation[,ekstrəpəu'leiʃən] *n*.外推法,推断
breed[bri:d] *n*.品种,种类	**mapping**['mæpiŋ] *n*.映射
instance['instəns] *n*.实例	**pervasive**[pə'veisiv] *adj*.普及的
jargon['dʒɑ:gən] *n*.行话	**propel**[prə'pel] *v*.推动,驱使
collie['kɔli] *n*.牧羊犬(一种源于苏格兰的高大聪敏长毛的牧羊狗)	**caveat**['keiviæt] *n*.警告
chihuahua[tʃi'wɑ:wɑ:] *n*.吉娃娃(一种产于墨西哥的狗)	

 Abbreviations

OOP	Object-Oriented Programming	面向对象的程序设计
PARC	Palo Alto Research Center	(施乐公司)帕洛阿尔托研究中心
UML	Unified Modeling Language	统一建模语言

Complex Sentences

1. **Original:** Computer programming methodology focuses on data rather than processes, with programs composed of self-sufficient modules (objects) containing all the information needed to manipulate a data structure.

   Translation: 计算机程序设计方法强调的是数据而不是过程，它使用由自给自足的模块（对象）组成的程序，这些模块包含了处理数据结构所需的所有信息。

2. **Original:** Some languages go further: Java uses the default access modifier to restrict access also to classes in the same package, C# and VB.NET reserve some members to classes in the same assembly using keywords internal (C#) or Friend (VB.NET), and Eiffel and C++ allow one to specify which classes may access any member.

   Translation: 一些语言可更进一步：Java使用默认访问控制修饰符来限制访问同一个包中的类，C#和VB.NET使用关键词internal (C#) 或Friend (VB.NET)为在同一个集合中的类保留了一些类的成员，而Eiffel和C++允许程序员指定哪一个类可以访问任何任意成员。

Exercises

I. Read the following statements carefully, and decide whether they are true (T) or false (F) according to the text.

___1. Object-oriented programming may be seen as a collection of cooperating objects having capabilities of receiving messages, processing data, and sending messages to each other.

___2. Object-oriented programming emerged primarily in the early 1990s.

___3. This article introduces nine fundamental concepts in the definitions of OOP in all.

___4. The rising popularity of GUIs and C++ enhanced the development of OOP as the dominant programming methodology.

___5. Object is a pattern (exemplar) of a class at runtime.

II. Choose the best answer to each of the following questions.

1. Which of the following is RIGHT about the fundamental concepts of OOP?

   (A) Class abstractly defines a thing's characteristics and the thing's behaviors.

   (B) Subclasses are specialized to restrict attributes and behaviors their parent classes have.

   (C) Polymorphism enables objects with different data types to respond to different type-specific method calls severally.

   (D) Multiple inheritance is an inheritance form in which more than one subclasses inherit from the same ancestor class.

2. Which of the following first explicitly introduced the term object-oriented programming?

   (A) Simula 67　　(B) Smalltalk　　(C) C++　　　　(D) Java

3. Which of the following is WRONG about the contrast between OOP and traditional programming?

(A) OOP enables non-programmers familiar with the problem domain to recognize the code more readily.

(B) As opposed to OOP, the traditional programming may be seen as a group of tasks to compute or a collection of subroutines.

(C) OOP can conceal the functional details of a class from objects that send messages to it , thereby can make changes more easily in future.

(D) Different languages primarily designed for OOP or traditional methodologies are incompatible with each other.

III. Translating.

1. Object-oriented programming was deployed in part as an attempt to address this problem by strongly emphasizing discrete units of programming logic and re-usability in software.

2. In the past decade Java has emerged in wide use partially because of its similarity to C and to C++, but perhaps more importantly because of its implementation using a virtual machine that is intended to run code unchanged on many different platforms.

≫ Section B   Introduction to the C# Language and the .NET Framework

*C# is an **elegant** and type-safe object-oriented language that enables developers to build a wide range of secure and robust applications that run on the .NET Framework.* You can use C# to create traditional Windows client applications, XML Web services, distributed components, client-server applications, database applications, and much, much more. Microsoft Visual C# 2005 provides an advanced code editor, convenient user interface designers, integrated debugger, and many other tools to facilitate rapid application development based on version 2.0 of the C# language and the .NET Framework.

⊖ C# Language

C# *syntax* is highly expressive with less than 90 keywords, which makes it simple and easy to learn. The *curly-brace* syntax of C# will be instantly recognizable to anyone familiar with C, C++ or Java. Developers who know any of these languages are typically able to begin working productively in C# within a very short time. C# syntax simplifies many of the complexities of C++ while providing powerful features such as nullable value types, *enumerations*, *delegates*, anonymous methods and direct memory access, which are not supported in Java. C# also supports generic methods and types, which provide increased type safety and performance, and iterators, which enable implementers of collection classes to define custom iteration behaviors that are simple to use by client code.

As an object-oriented language, C# supports the concepts of encapsulation, inheritance and polymorphism. All variables and methods, including the Main method, the application's entry point, are encapsulated within class definitions. A class may inherit directly from one parent class, but it may implement any number of interfaces. Methods that *override* virtual methods in a parent class require the override keyword as a way to avoid accidental redefinition. In C#, a struct is like a lightweight class; it is a stack-allocated type that can implement interfaces but does not support inheritance.

In addition to these basic object-oriented principles, C# facilitates the development of software components through several innovative language constructs, including:

- Encapsulated method signatures called delegates, which enable type-safe event notifications.
- Properties, which serve as accessors for private member variables.
- Attributes, which provide *declarative* metadata about types at run time.
- *Inline* XML documentation comments.

If you need to interact with other Windows software such as **COM** objects or *native* Win32 **DLLs**, you can do this in C# through a process called "Interop." Interop enables C# programs to do just about anything that a native C++ application can do. C# even supports pointers and the concept of "unsafe" code for those cases in which direct memory access is absolutely critical.

The C# build process is simple compared to C and C++ and more flexible than that in Java. There are no separate header files, and no requirement that methods and types be declared in a particular order. A C# source file may define any number of classes, structs, interfaces, and events.

.NET Framework Platform Architecture

C# programs run on the .NET Framework, an integral component of Windows that includes a virtual execution system called the common language runtime (CLR) and a unified set of class libraries. *The CLR is Microsoft's commercial implementation of the common language infrastructure (CLI), an international standard that is the basis for creating execution and development environments in which languages and libraries work together seamlessly.*

Source code written in C# is compiled into an intermediate language (**IL**) that conforms to the CLI specification. *The IL code, along with resources such as bitmaps and strings, is stored on disk in an executable file called an assembly, typically with an extension of .exe or .dll.* An assembly contains a manifest that provides information on the assembly's types, version, culture, and security requirements.

When the C# program is executed, the assembly is loaded into the CLR, which might take various actions based on the information in the manifest. Then, if the security requirements are met, the CLR performs just in time (**JIT**) compilation to convert the IL code into native machine instructions. The CLR also provides other services related to automatic garbage collection, exception handling, and resource management. Code that is executed by the CLR is sometimes referred to as "*managed code*," in contrast to "*unmanaged code*" which is compiled into native machine language that targets a specific system. The following diagram illustrates the compile-time and run time relationships of C# source code files, the base class libraries, assemblies, and the CLR. See the Figure 5-1.

Language interoperability is a key feature of the .NET Framework. Because the IL code produced by the C# compiler conforms to the Common Type Specification (**CTS**), IL code generated from C# can interact with code that was generated from the .NET versions of Visual Basic, Visual C++, Visual J#, or any of more than 20 other CTS-compliant languages. A single assembly may contain multiple modules written in different .NET languages, and the types can reference each other just as if they were written in the same language.

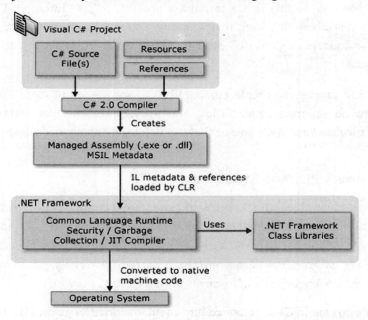

Figure 5-1   Compiling and running relationships

In addition to the run time services, the .NET Framework also includes an extensive library *of over 4000 classes organized into namespaces that provide a wide variety of useful functionality for everything from file input and output to string manipulation to XML **parsing**, to Windows Forms controls*. The typical C# application uses the .NET Framework class library extensively to handle common "**plumbing**" **chores**.

Words

elegant['eligənt] *adj*.简洁的，典雅的	**inline**[inlain] *adj*.内嵌的
syntax['sintæks] *n*.语法	**native**['neitiv] *adj*.标准的，专属的
enumeration[i,nju:mə'reiʃən] *n*.枚举	**parse**[pɑːz] *v*.解析
delegate['deligit] *n*.委托	**plumbing**[plʌmiŋ] *adj*.了解的，查明的，测量的
override[,əuvə'raid] *v*.优先于，压倒，使无效	**chore**[tʃɔː] *n*.日常工作，例行事务
declarative[di'klærətiv] *adj*.说明的，陈述的，公布的	

Phrases

type-safe	类型安全	**managed code**	托管代码
curly-brace	卷曲的大括号	**unmanaged code**	非托管代码

Abbreviations

COM	Component Object Model	组件对象模型
DLL	Dynamic Link Library	动态链接库
IL	Intermediate Language	中间语言
JIT	Just In Time	实时，激活
CTS	Common Type Specification	通用类型说明

 Complex Sentences

1. **Original:** The CLR is Microsoft's commercial implementation of the common language infrastructure (CLI), an international standard that is the basis for creating execution and development environments in which languages and libraries work together seamlessly.

   Translation: 通用语言运行时（CLR）是微软公司对通用语言基础结构（CLI）的商业实现，它是一个国际标准，该标准是创建语言和库无缝协调工作的执行和开发环境的基础。

2. **Original:** In addition to the run time services, the .NET Framework also includes an extensive library of over 4000 classes organized into namespaces that provide a wide variety of useful functionality for everything from file input and output to string manipulation to XML parsing, to Windows Forms controls.

   Translation: 除了运行时服务，.NET Framework还包括一个巨大的类库，有超过4000个被组织到命名空间中的类。这个类库提供了多种有用的功能，从文件输入输出到串操作，到XML解析，再到Windows窗体控制。

 **Exercises**

Cmprehension of the Text

I. Read the following statements carefully, and decide whether they are true (T) or false (F) according to the text.

___1. C# prohibits direct memory access because those cases are absolutely critical.

___2. A C# source file may define any number of classes, structs, interfaces, and events.

___3. A C# struct is like a lightweight class that can implement interfaces and support inheritance.

___4. "Unmanaged code" is compiled into native machine language that targets a specific system, not executed by the CLR.

___5. A single assembly may contain multiple modules written in different .NET languages.

II. Choose the best answer to each of the following questions.

1. Which of the following is WRONG about the C# syntax?

(A) All variables and methods are encapsulated within class definitions in C#.

(B) C# provides nullable value types, enumerations, delegates, anonymous methods and direct memory access.

(C) To avoid accidental redefinition, methods that override virtual methods in a parent class require overriding name in C#.

(D) There is no separate header files, and no requirement that methods and types be declared in a particular order in C#.

2. Which of the following is NOT included in the functions of CLR?

(A) Performs JIT compilation to convert the IL code into native machine instructions

(B) Automatic garbage collection

(C) Exception handling and resource management

(D) Includes an extensive class library that provides a wide variety of useful functionality

3. Which of the following is the RIGHT code conversion sequence in a complete execution process of C# program?

(A) Source code, IL, native machine code

(B) Source code, class libraries, native machine code

(C) Source code, assembly, class libraries

(D) Source code, CLR, native machine instructions

III. Translating.

1. C# is an elegant and type-safe object-oriented language that enables developers to build a wide range of secure and robust applications that run on the .NET Framework.

2. The IL code, along with resources such as bitmaps and strings, is stored on disk in an executable file called an assembly, typically with an extension of .exe or .dll.

Part 2　**Simulated Writing: Outline**

⊝ Introduction

An outline is a general plan of the material that is to be presented in a speech or a paper. The outline shows the order of the various topics, the relative importance of each, and the relationship between the various parts.

⊝ Order in an Outline

There are many ways to arrange the different parts of a subject. Sometimes, a chronological arrangement works well. At other times, a spatial arrangement is best suited to the material. The most common order in outlines is to go from the general to the specific. This means you begin with a general idea and then support it with specific examples.

⊝ Thesis Statement of Summarizing Sentence

All outlines should begin with a thesis statement or summarizing sentence. This thesis sentence presents the central idea of the paper. It must always be a complete, grammatical sentence, specific and brief, which expresses the point of view you are taking towards the subject.

⊝ Types of Outlines

The two main types of outlines are the topic outline and the sentence outline. In the topic outline, the headings are given in single words or brief phrases. In the sentence outline, all the headings are expressed in complete sentences.

⊝ Format

A focused paper is the key to writing a good essay. Here's the format for the outline.

Topic / Title

Thesis: Summarizing sentence

I. Introduction of Essay

　A. Write a few sentences that lead into the main point of your essay

　B. End the paragraph with your thesis statement (3 main points you are going to support)

　　1. First point in thesis

　　2. Second point in thesis

　　3. Third point in thesis

II. Body of Essay - (Should have at least three paragraphs)

　A. Topic One - First Point in Thesis - (1st paragraph of Body)

        1. Support your point with either quotations or solid evidence

        2. Have at least five sentences

    B. Topic Two - Second Point in Thesis - (2nd paragraph of Body)

        1. Support your point with either quotations or solid evidence

        2. Have at least five sentences

    C. Topic Three - Third Point in Thesis - (3rd paragraph of Body)

        1. Support your point with either quotations or solid evidence

        2. Have at least five sentences

III. Conclusion

    A. Write a few sentences summarizing your essay

    B. Restate your thesis and how you proved your point

Once you fill in the blanks to this outline with your topic and information, the rest is easy! Make sure to check your spelling and punctuation, and then you're good to go!

Sample

A. Topic Outline

Choices in College and After

Thesis: The decisions I have to make in choosing college courses, depends on larger questions I am beginning to ask myself about my life's work.

I. Two decisions described

  A. Art history or chemistry

    1. Professional considerations

    2. Personal considerations

  B. A third year of French?

    1. Practical advantages of knowing a foreign language

    2. Intellectual advantages

    3. The issue of necessity

II. Definition of the problem

  A. Decisions about occupation

  B. Decisions about a kind of life to lead

III. Temporary resolution of the problem

  A. To hold a professional possibility: chemistry

  B. To take advantage of cultural gains already made: French

B. Sentence Outline

Choices in College and After

Thesis: The decisions I have to make in choosing college courses, depends on larger questions I am beginning to ask myself about my life's work.

I. I have two decisions to make with respect to choosing college courses in the immediate future.

    A. One is whether to elect a course in art history or in chemistry.

      1. One time in my life, I planned to be a chemical engineer professionally.

      2. On the other hand, I enjoy art and plan to travel to see more of it.

    B. The second decision is whether to continue a third year of French beyond the basic college requirement.

      1. French might be useful both in engineering and travelling.

      2. Furthermore, I am eager to read good books which are written in French.

      3. How necessary are these considerations in the light of other courses I might take instead?

II. My problem can be put in the form of a dilemma involving larger questions about my whole future.

    A. On the one hand I want to hold a highly-trained position in a lucrative profession.

    B. On the other hand I want to lead a certain kind of life, with capacities for values not connected with the making of money.

III. I will have to make a decision balancing the conflicting needs I have described.

    A. I will hold professional possibilities by electing chemistry.

    B. I will improve and solidify what cultural proficiency in another language I have already gained, by electing French.

Part 3   Listening and Speaking

≫ Dialogue: Getting to Know Java Runtime Environment (JRE) and Java Virtual Machine (JVM)

*(Before the first lesson of Java programming, Mark downloaded a simple **applet** for a sample to learn, but he found that it couldn't run.)*

Mark	Excuse me, Henry and Sophie. Could you help me?
Henry	Sure. What's wrong?

Mark	Why can't this applet run? Its source code is correct.
Sophie	Has your computer installed Java Runtime Environment[1]?
Mark	Not yet. What's Java Runtime Environment?
Henry	JRE *for short*, it is a software platform from Sun Microsystems that allows a computer to run applets and applications written in the Java programming language. [2]
Sophie	It contains Java Virtual Machine, Java libraries and some other components.
Mark	What is Java Virtual Machine used for?
Sophie	A Java Virtual Machine, JVM for short, is a set of computer software programs and data structures which use a virtual machine model for the execution of other computer programs and scripts. It hides the details of the computer hardware on which their programs run. [3]
Mark	What's the relationship between JVM and JRE? [4]
Henry	The JVM, which is the instance of the JRE, *comes into action* when a Java program is executed. When execution is complete, this instance is garbage- collected. The JVM is distributed along with a set of standard class libraries which implement the Java API. The virtual machine and API have to be consistent with each other and are therefore bundled together as the JRE[5].
Sophie	So, this can be considered a virtual computer in which the virtual machine is the processor and the API is the user interface.
Henry	The JVM is a crucial component of the Java Platform. The use of the same **bytecode** for all platforms allows Java to be described as "compile once, run anywhere".
Mark	Bytecode?
Henry	Yes. JVM operates on a form of computer intermediate language commonly referred to as Java bytecode, which is normally but not necessarily generated from Java source code. Programs intended to run on a JVM must be compiled into this standardized portable binary format. [6]
Mark	Is it visible for us or just an automatic temporary code?
Sophie	It typically comes in the form of .class files.
Mark	But a large program may consist of many classes in different files.

Sophie	Yes. For easier distribution of large programs, multiple class files may be packaged together in a .jar file, short for Java archive.
Henry	The JVM runtime executes .class or .jar files, emulating the JVM instruction set by interpreting it, or using a JIT compiler, *shortened form* of just-in-time compiler, such as Sun's HotSpot.
Sophie	JIT? I've heard a little of it before. What technology does JIT use?
Henry	JIT compiles parts of the bytecode that have similar functionality at the same time, and hence reduces the amount of time needed for compilation. It is used in most JVMs today to achieve greater speed. [7]
Mark	So much knowledge I'm interested in! I'll do my best to learn. Thank you very much!

Exercises

Work in pairs, and make up a similar conversation by replacing the scenarios with other material in the below.

Getting to Know.NET Framework and Common Language Runtime (CLR)

[1] The Microsoft .NET Framework

[2] .NET Framework is a software technology that is available with several Microsoft Windows operating systems. The .NET Framework is a key Microsoft offering and is intended to be used by most new applications created for the Windows platform. It includes a large library of pre-coded solutions to common programming problems and a virtual machine that manages the execution of programs written specifically for the framework.

[3] The Common Language Runtime (CLR) is the virtual machine component of Microsoft's .NET initiative. The CLR provides the appearance of an application virtual machine so that programmers need not consider the capabilities of the specific CPU that will execute the program.

[4] CLR and .Net Framework

[5] Programs written for the .NET Framework execute in the CLR. The class library and the CLR together compose the .NET Framework.

[6] The CLR runs a form of bytecode called the Common Intermediate Language (**CIL**) which defines an execution environment for program code.

[7] On .NET the byte-code is always compiled before execution, either Just In Time (JIT) or in

advance of execution using the utility. During compiling time, a .NET compiler converts source code into CIL code. At runtime, the CLR's just-in-time compiler converts the CIL code into code native to the operating system. Alternatively, the CIL code can be compiled to native code in a separate step prior to runtime. This speeds up all later runs of the software as the CIL-to-native compilation is no longer necessary.

 Words

applet [eplet] *n*.Java小应用程序（application let）	**bytecode** ['baitkəud] *adj*.字节码

 Phrases

for short	简称，缩写	**shortened form**	简称，简写
come into action	起作用，投入战斗		

 Abbreviations

JRE	Java Runtime Environment	Java运行环境
JVM	Java Virtual Machine	Java虚拟机
CIL	Common Intermediate Language	通用中间语言

>> Listening Comprehension: IDE

Listen to the passage and the following 3 questions based on it. After you hear a question, there will be a break of 10 seconds. During the break, you will decide which one is the best answer among the four choices marked (A), (B), (C) and (D).

Questions

1. What is the correct full name of the abbreviation "IDE" in this article?
2. What is the greatest benefit brought by IDE for software developers according to this article?
3. Which of the following items is NOT integrated in the IDE firstly used in Dartmouth BASIC?

Choices

1. (A) Interface Development Environment   (B) Integrated Development Environment
   (C) Integrated Development Editor   (D) Interface Debugging Editor
2. (A) Learning a language   (B) Increasing developing productivity
   (C) Piecing together command lines   (D) Compiling code
3. (A) File management   (B) Compilation
   (C) Debugging   (D) Graphical user interface

Words

abbreviate[ə'bri:vieit] *v.*缩写，简写	**cohesive**[kəu'hi:siv] *adj.*粘着的
facility[fə'siliti] *n.*工具，便利	**flowchart**[fləu'tʃɑ:t] *n.*流程图
interpreter[in'tə:pritə] *n.*解释程序	**keypunch**['ki:pʌntʃ] *n.*键盘穿孔机
configuration[kən,figju'reiʃən] *n.*配置	

▶▶ Dictation: Ada Lovelace, the First Programmer

This passage will be played THREE times. Listen carefully, and fill in the blanks with the words you have heard.

Ada Byron, Lady Lovelace, was one of the most *picturesque* ____1____ in computer history. Daughter of the Romantic ____2____ Lord Byron, Lady Ada Lovelace was known as the "*enchantress* of numbers" who ____3____ with Charles Babbage, the ____4____ of the first mechanical thinking/calculating machine.

It was at a dinner party that Ada ____5____ in November, 1834, Babbage's ____6____ for a new calculating ____7____ , the Analytical Engine. He *conjectured*: what if a calculating engine could not only foresee but could ____8____ on that foresight. Ada was ____9____ by the "universality of his ideas". ____10____ anyone else was.

Ada ____11____ a paper on Babbage's Engines by General Menabrea, later to be prime minister of the newly united Italy. When she showed Babbage her translation he suggested that she ____12____ her own notes, which ____13____ to be three times the ____14____ of the original article. Letters between Babbage and Ada flew back and forth ____15____ fact and *fantasy*. In her article, ____16____ in 1843, Lady Lovelace *anticipated* the development of computer software, ____17____ and computer music.

Ada Lovelace *devised* a method of using *punch* ____18____ to calculate Bernoulli numbers, becoming the first computer ____19____. In her honor, the U.S. Department of Defense ____20____ its computer language "Ada" in 1980.

Words

picturesque[,piktʃə'resk] *adj.*独特的	**anticipate**[æn'tisipeit] *v.*预期，预见
enchantress[in'tʃɑ:ntris] *n.*女巫	**devise**[di'vaiz] *v.*设计
conjecture[kən'dʒektʃə] *v.*推测，猜想	**punch**[pʌntʃ] *n.*打孔机
fantasy['fæntəsi] *n.*幻想，狂想，想象	

Databases

Part 1   Reading and Translating

>>> Section A   Distributed Database

Introduction

A distributed database is a database that is under the control of a central database management system (DBMS) in which storage devices are not all attached to a common CPU. It may be stored in multiple computers located in the same physical location, or may be dispersed over a network of interconnected computers.

Collections of data (e.g. in a database) can be distributed across multiple physical locations. A distributed database is distributed into separate *partitions/ fragments*. Each partition/fragment of a distributed database may be *replicated* (i.e. *redundant fail-overs*, **RAID** like).

Besides distributed database replication and fragmentation, there are many other distributed database design technologies. For example, local *autonomy*, synchronous and *asynchronous* distributed database technologies.

These technologies' implementation can and does depend on the needs of the business and the sensitivity/confidentiality of the data to be stored in the database, and hence the price the business is willing to spend on ensuring data security, consistency and integrity.

Basic Architecture

A database users access the distributed database through:

- Local applications — applications which do not require data from other sites.
- Global applications — applications which do require data from other sites.

Important Considerations

Care with a distributed database must be taken to ensure the following:

- The distribution is *transparent* — users must be able to interact with the system as if it were one logical system. This applies to the system's performance, and methods of access amongst other things.
- Transactions are transparent — each transaction must maintain database integrity across multiple databases. Transactions must also be divided into subtransactions, each subtransaction affecting one database system.

Advantages of Distributed Databases

- Reflects organizational structure — database fragments are located in the departments they relate to.
- Local autonomy — a department can control the data about them (as they are the ones familiar with it.)
- Improved availability — a fault in one database system will only affect one fragment, instead of the entire database.
- Improved performance — *data is located near the site of greatest demand, and the database systems themselves are parallelized, allowing load on the databases to be balanced among servers.* (A high load on one module of the database won't affect other modules of the database in a distributed database.)
- Economics — it costs less to create a network of smaller computers with the power of a single large computer.
- Modularity — systems can be modified, added and removed from the distributed database without affecting other modules (systems).

Disadvantages of Distributed Databases

- Complexity — extra work must be done by the **DBAs** to ensure that the distributed nature of the system is transparent. Extra work must also be done to maintain multiple *disparate* systems, instead of a big one. Extra database design work must also be done to *account for* the disconnected nature of the database — for example, joins become *prohibitively* expensive when performed across multiple systems.

- Economics — increased complexity and a more extensive infrastructure means extra labor costs.
- Security — remote database fragments must be secured, and they are not centralized so the remote sites must be secured as well. The infrastructure must also be secured (e.g., by *encrypting* the network links between remote sites).
- Difficult to maintain integrity — in a distributed database, enforcing integrity over a network may require too much of the network's resources to be feasible.
- Inexperience — distributed databases are difficult to work with, and as a young field there is not much readily available experience on proper practice.
- Lack of standards – there are no tools or methodologies yet to help users convert a centralized DBMS into a distributed DBMS.
- Database design more complex – besides of the normal difficulty, the design of a distributed database has to consider fragmentation of data, allocation of fragments to specific sites and data replication.

Distributed Database Management System

A distributed database management system is a software system that permits the management of a distributed database and makes the distribution transparent to the users. A distributed database is a collection of multiple, logically interrelated databases distributed over a computer network. Sometimes "distributed database system" is used to refer jointly to the distributed database and the distributed DBMS.

 Words

partition[pɑːˈtiʃ ən] *n.*分区，部分	**transparent**[trænsˈpɛərənt] *adj.*透明的
fragment[ˈfrægmənt] *n.*片段，分段	**disparate**[ˈdispərit] *adj.*异类的，完全不同的
replicate[ˈreplikit] *v.*复制	**prohibitively**[prəˈhibitivli] *adv.*高得惊人地，抑制
redundant[riˈdʌndənt] *adj.*冗余的	购买地
autonomy[ɔːˈtɔnəmi] *n.*自治权，自主权	**encrypt**[inˈkript] *v.*加密
asynchronous[eiˈsiŋkrənəs] *adj.*不同时的，异步的	

Phrases

fail-over	故障备份替换	**account for**	说明，解释

Abbreviations

DBMS	DataBase Management System	数据库管理系统
RAID	Redundant Array of Independent Disk	独立冗余磁盘阵列
DBA	Database Administrator	数据库管理员

Complex Sentences

1. **Original**: These technologies' implementation can and does depend on the needs of the business and the sensitivity/confidentiality of the data to be stored in the database, and hence the price the business is willing to spend on ensuring data security, consistency and integrity.

   Translation: 这些技术的实现能够并且确实取决于企业的需求，同时还取决于要存储在数据库中的数据的敏感性/机密性，因此也取决于企业愿意花费在保证数据安全、一致性和完整性上的代价。

Exercises

I. Read the following statements carefully, and decide whether they are true (T) or false (F) according to the text.

___1. A distributed database is a database stored in multiple computers located in the same physical location.

___2. Users access the distributed database through local applications and global applications.

___3. A distributed database is composed of a collection of logically interrelated but physically dispersed databases.

___4. Each partition/fragment of a distributed database stores different data non-repeatedly.

___5. So far, distributed database is a young field without much readily available experience on proper practice.

II. Choose the best answer to each of the following questions.

1. Which of the following is WRONG about the meaning of the word "transparent" in the article?

   (A) Users interact with the distributed database as if it were one logical system.

   (B) Users implement transactions in the distributed database as if they were in one logical system.

   (C) Users may access data in the distributed database with no need to know where the data is located physically.

   (D) Users work on the distributed database as if it did not exist.

2. Which of the following is NOT mentioned in this article about the distributed database design technologies?

   (A) Replication and fragmentation   (B) Local autonomy

   (C) Synchronous and asynchronous   (D) Fault-tolerant

3. Which of the following is NOT the drawback of distributed database compared to the traditional centralized database?
(A) Experience on proper practice   (B) Performance
(C) Design complexity   (D) Integrity maintenance.

III. Translating.

1. A distributed database is a database that is under the control of a central database management system (DBMS) in which storage devices are not all attached to a common CPU.
2. Data is located near the site of greatest demand, and the database systems themselves are parallelized, allowing load on the databases to be balanced among servers.

>>> Section B   Microsoft SQL Server

Microsoft SQL Server is a relational database management system (**RDBMS**) produced by Microsoft. Its primary query language is **Transact-SQL**, an implementation of the ANSI/ISO standard Structured Query Language (SQL) used by both Microsoft and Sybase.

Architecture

The architecture of Microsoft SQL Server is broadly divided into three components: **SQLOS** which implements the basic services required by SQL Server, including thread scheduling, memory management and I/O management; the Relational Engine, which implements the relational database components including support for databases, tables, queries and *stored procedures* as well as implementing the type system; and the **Protocol** Layer which exposes the SQL Server functionality.

SQLOS

SQLOS is the base component in the SQL Server architecture. It implements functions normally associated with the operating system—thread scheduling, memory management, I/O management, buffer pool management, resource management, synchronization *primitives* and locking, and *deadlock* detection. Because the requirement of SQL Server is highly specialized, it implements its own memory and thread management system, rather than using the generic one already implemented in the operating system. It divides all the operations it performs into a series of tasks such as background maintenance *jobs* and processing requests from clients. *Internally, a pool of worker threads is maintained, onto which the tasks are scheduled. A task is associated with the thread until it is completed, only after its completion is the thread freed and returned to the pool. If there are no free threads to assign the task to, the task is temporarily blocked.* Each worker thread is mapped onto either an operating system thread or a *fiber*. Fibers are user mode threads that implement co-operative multitasking. Using fibers means SQLOS does all the book-keeping of thread management itself, but then it can optimize them for its particular use. SQLOS also includes synchronization primitives for locking as well as monitoring for the worker threads to detect and recover from deadlocks.

SQLOS handles the memory requirements of SQL Server as well. Reducing disc I/O is one of the primary goals of specialized memory management in SQL Server. It maintains a buffer pool, which is used to cache data pages from the disc, and to satisfy the memory requirements for the query processor, and for other internal data structures. *SQLOS monitors all the memory allocated from the buffer pool, ensuring that the components return unused memory to the pool, and* **shuffles** *data out of the cache to make room for newer data. For changes that are made to the data in buffer, SQLOS writes the data back to the disc lazily, that is when the disc subsystem is either free, or there have been significant numbers of changes made to the cache, while still serving requests from the cache.* For this, it implements a Lazy Writer, which handles the task of writing the data back to persistent storage.

SQL Server normally supports up to 2 GB memory on x86 hardware, though it can be configured to use up to 64 GB if the Address Windowing Extension is used in the supporting operating system. For x64 hardware, it supports 8 TB of memory, and 7 TB for **IA**-64 systems (currently it is limited by Windows Server 2003 **SP**1 to 1TB). However, when running x86 versions of SQL Server on x64 hardware, it can access 4 GB of memory without any special configuration.

Relational Engine

The Relational Engine implements the relational data store using the capabilities provided by SQLOS, which is exposed to this layer via the private SQLOS API. It implements the type system, to define the types of the data that can be stored in the tables, as well as the different types of data items (such as tables, indexes, logs etc) that can be stored. It includes the Storage Engine, which handles the way data is stored on persistent storage devices and provides methods for fast access to the data. The storage engine implements log-based transaction to ensure that any changes to the data are **ACID** compliant. It also includes the query processor, which is the component that retrieves data. *SQL queries specify what data to retrieve, and the query processor optimizes and translates the query into the sequence of operations needed to retrieve the data.* The operations are then performed by worker threads, which are scheduled for execution by SQLOS.

Protocol Layer

Protocol Layer implements the external interface to SQL Server. All operations that can be **invoked** on SQL Server are communicated to it via a Microsoft-defined format, called Tabular Data Stream (**TDS**). TDS is an application layer protocol, used to transfer data between a database server and a client. Initially designed and developed by Sybase Inc. for their Sybase SQL Server relational database engine in 1984, and later by Microsoft in Microsoft SQL Server, TDS packets can be *encased* in other physical transport dependent protocols, including **TCP/IP,** named pipes, and shared memory. Consequently, access to SQL Server is available over these protocols. In addition, the SQL Server API is also exposed over Web services.

History

The code base for MS SQL Server (prior to version 7.0) originated in Sybase SQL Server, and was Microsoft's entry to the enterprise-level database market, competing against Oracle, IBM, and, later, Sybase itself. Microsoft, Sybase and Ashton-Tate originally teamed up to create and market the first version named SQL Server 1.0 for OS/2 (about 1989) which was essentially the same as Sybase SQL Server 3.0 on Unix, **VMS,** etc. Microsoft SQL Server 4.2 was shipped around 1992 (available bundled with Microsoft OS/2 version 1.3). Later Microsoft SQL Server 4.21 for Windows NT was released at the same time as Windows NT 3.1. Microsoft SQL Server v6.0 was the first version designed for NT, and did not include any direction from Sybase.

About the time Windows NT was released, Sybase and Microsoft *parted* ways and pursued their own design and marketing schemes. Microsoft negotiated exclusive rights to all versions of SQL Server written for Microsoft operating systems. Later, Sybase changed the name of its product to Adaptive Server Enterprise to avoid confusion with Microsoft SQL Server. Until 1994 Microsoft's SQL Server carried three Sybase copyright notices as an indication of its origin.

Since parting ways, several revisions have been done independently. SQL Server 7.0 was the first true GUI based database server and was a rewrite from the legacy Sybase code. It was succeeded by SQL Server 2000, which was the first edition to be launched in a variant for the IA-64 architecture.

In the next six years since release of Microsoft's previous SQL Server product (SQL Server 2000), advancements have been made in performance, the client IDE tools, and several complementary systems that are packaged with SQL Server 2005. These include: an **ETL** tool (SQL Server Integration Services or **SSIS**), a Reporting Server, an **OLAP** and data mining server (Analysis Services), and several messaging technologies, specifically Service Broker and Notification Services.

SQL Server 2008

The current version of SQL Server, SQL Server 2008, (code-named "Katmai",) was released on August 6, 2008 and aims to make data management *self-tuning*, self organizing, and self maintaining with the development of SQL Server Always On technologies, to provide near-zero downtime. SQL Server 2008 will also include support for structured and semi-structured data, including digital media formats for pictures, audio, video and other multimedia data. In current versions, such multimedia data can be stored as **BLOB**s (binary large objects), but they are generic bit streams. *Intrinsic* awareness of multimedia data will allow specialized functions to be performed on them. According to Paul Flessner, senior Vice President, Server Applications, Microsoft Corp., SQL Server 2008 can be a data storage *back-end* for different varieties of data: XML, email, time/calendar, file, document, *spatial,* etc as well as perform search, query, analysis, sharing, and synchronization across all data types.

Other new data types include specialized date and time types and a spatial data type for location-dependent data. Better support for unstructured and semi-structured data is provided using the FILESTREAM data type which can be used to reference any file stored on the file system. Structured data and metadata about the file is stored in SQL Server database, whereas the unstructured component is stored in the file system. Such files can be accessed both via Win32 file handling APIs and via SQL Server using T-SQL; doing the latter accesses the file data as a binary BLOB. *Backing up* and restoring the database backs up or restores the referenced files as well. SQL Server 2008 also natively supports *hierarchical* data, and includes T-SQL constructs to directly deal with them, without using *recursive* queries.

The Full-Text Search functionality has been integrated with the database engine, which simplifies management and improves performance.

 Words

protocol[ˈprəutəkɔl] *n.*协议	**encase**[inˈkeis] *v.*嵌入，装入
primitive[ˈprimitiv] *n.*原语	**part**[pɑːt] *v.*分开，分离
deadlock[ˈdedlɔk] *n.*死锁	**intrinsic**[inˈtrinsik] *adj.*固有的，内在的
job[dʒɔb] *n.*作业	**spatial**[ˈspeiʃəl] *adj.*空间的
fiber[ˈfaibə] *n.*纤程	**hierarchical**[ˌhaiəˈrɑːkikəl] *adj.*分等级的
shuffle[ˈʃʌfl] *v.*搬移	**recursive**[riˈkəːsiv] *adj.*递归的
invoke[inˈvəuk] *v.*调用	

 Phrases

stored procedure	存储过程	**back-end**	后端
self-tuning	自校正	**back up**	备份

 Abbreviations

RDBMS	Relational DataBase Management System	关系型数据库管理系统

T-SQL	Transact-SQL	事务型SQL
SQLOS	SQL Operating System	SQL操作系统
IA	Intel Architecture	英特尔体系结构
SP	Service Pack	服务包，补丁
ACID	Atomicity, Consistency, Isolation and Durability	原子性、一致性、隔离性、持久性
TDS	Tabular Data Stream	表格数据流
TCP/IP	Transmission Control Protocol/Internet Protocol	传输控制协议/因特网协议
VMS	Virtual Memory System	虚拟内存系统（一种操作系统）
ETL	Extraction-Transformation-Loading	数据抽取、转换和加载
SSIS	SQL Server Integration Services	SQL Server集成服务
OLAP	On-Line Analysis Processing	联机分析处理
BLOB	Binary Large OBject	二进制大对象

Complex Sentences

1. **Original:** Internally, a pool of worker threads is maintained, onto which the tasks are scheduled. A task is associated with the thread until it is completed, only after its completion is the thread freed and returned to the pool. If there are no free threads to assign the task to, the task is temporarily blocked.

   Translation: 在SQLOS内部，维护着一个工作者线程池，任务在这个池上被调度。一个任务与线程相结合直到该任务完成，只有在其完成之后线程才被释放并返回到池中。如果没有空闲线程分配给某任务，该任务就被临时阻塞。

2. **Original:** SQLOS monitors all the memory allocated from the buffer pool, ensuring that the components return unused memory to the pool, and shuffles data out of the cache to make room for newer data.

   Translation: SQLOS监视所有从缓冲池分配的内存，确保组件将不用的内存归还给缓冲池，并将数据移出高速缓存，为新数据腾出空间。

Exercises

I. **Read the following statements carefully, and decide whether they are true (T) or false (F) according to the text.**

____1. The architecture of Microsoft SQL Server is composed of three components.

____2. Access to SQL Server is only available over TDS protocol in the Protocol Layer.

____3. The code base for the prime versions of MS SQL Server adopted a direction from Sybase.

____4. Both Microsoft and Sybase used Transact-SQL as the primary query language of their SQL Server products.

____5. In MS SQL Server 2008, structured data, metadata and the unstructured component of a file stored on the file system are all stored in SQL Server database.

II. Choose the best answer to each of the following questions.

1. Which of the following is responsible for implementing log-based transaction in MS SQL Sever?

   (A) The storage engine   (B) Lazy Writer   (C) The query processor   (D) TDS

2. Which of the following is WRONG about the new features of MS SQL Sever 2008?

   (A) Support specialized date and time types and a spatial data type

   (B) Support multimedia data to be stored as bit stream type

   (C) Support T-SQL constructs to directly deal with hierarchical data

   (D) Support Full-Text Search functionality

3. Which of the following is RIGHT about the relationship between MS SQL Server and Sybase SQL Server?

   (A) Both of them are created and marketed by the cooperation between Microsoft and Sybase all the time.

   (B) They are independently developed by either Microsoft or Sybase from the origin.

   (C) Sybase has several copyrights to SQL Server at the beginning.

   (D) Microsoft has exclusive rights to all versions of SQL Server at the beginning.

III. Translating.

1. For changes that are made to the data in buffer, SQLOS writes the data back to the disc lazily, that is when the disc subsystem is either free, or there have been significant numbers of changes made to the cache, while still serving requests from the cache.

2. SQL queries specify what data to retrieve, and the query processor optimizes and translates the query into the sequence of operations needed to retrieve the data.

Part 2 Simulated Writing: Summary

➡Introduction

A summary is a condensed form of a document that represents its full sense. It should restate the author's main point, purpose, intent and supporting details in your own words and contain only the author's views. A effective summary can demonstrate understanding to someone in a senior position, allow readers to decide whether they wish to read the full document, provide an introduction of the document to those who don't have time to read it, and provide yourself with a better understanding of the document.

➡Procedure for Report Writing

Read the article carefully. Determine its structure. Identify the author's purpose in writing. To write a good summary it is important to thoroughly understand the material you are working with.

Reread, label, and underline. Look for section headings, bolded or italicized words, and subheadings. Divide the article into sections or stages of thought. Label, on the passage itself, each section or stage of thought. Underline key ideas and terms.

Write one-sentence summary, **on a separate sheet of paper, of each stage of thought**. This should be a brief outline of the article.

Write a thesis—a one-sentence—summary of the entire article. The thesis should express the central idea of the article, as you have determined it from the preceding steps. You may find it useful to keep in mind the information contained in the lead sentence or paragraph of most newspaper stories—the what, who, why, where, when, and how of the matter. For persuasive passages, summarize in a sentence the author's conclusion. For descriptive passages, indicate the subject of the description and its key features. Note: In some cases a suitable thesis may already be in the original passage. If so, you can quote it directly in your summary.

Write the first draft of your summary by combine the information from the first four steps into paragraphs. In either case, eliminate repetition. Eliminate less important information. Disregard minor details, or generalize them. Use as few words as possible to convey the main ideas.

Check your summary against the original passage, and make whatever adjustments are necessary for accuracy and completeness.

Revise your summary, inserting transitional words and phrases where necessary to ensure coherence. Check for style. Avoid series of short, choppy sentences. Combine sentences for a smooth, logical flow of ideas. Check for grammatical correctness, punctuation, and spelling.

Other Useful Information

- Cover the original as a whole and include any important ideas, data and conclusions.
- Use a simple organization: (1) main points; (2) main results; (3) conclusions/recommend-dations.
- Follow the original organization where possible and use the author's key words.
- Use 3rd person and present tense.
- Present in your own words and quote the author sparingly, if at all.
- Do not include anything that does not appear in the original, e.g. your own opinion.
- Make the summary clear and understandable and stand on its own.
- Be concise and brief and keep summary short: 3 to 7 sentences.

→**Sample**

Global Implications of Patent Law Variation

A patent is an exclusive right to use an invention for a certain period of time, which is given to an inventor as compensation for disclosure of an invention. Although it would be beneficial for the world economy to have uniform patent laws, each country has its own laws designed to protect domestic inventions and safeguard technology. Despite widespread variation, patent laws generally fall under one of two principles: the first-to-file and first-to-invent. The first-to-file principle awards a patent to the person or institution that applies for a patent first, while the first-to-invent principle grants the patent to the person or institution that was first to invent it. Most countries have adopted the first-to-file system. However, the United States maintains a first-to-invent system, despite obvious shortcomings. A result of countries employing different patent ownership is not recognized globally. On the contrary, ownership may change depending on the country. It is not uncommon for an invention to have two patent owners—one in the United States and one in the rest of the world. This unclear ownership often has economic consequences. If a company is interested in using a patented invention, it may be unable to receive permission from both patent owners, which in turn may prevent manufacture of a particular product. Even if permission is received from both owners, pay royalties to both may be quite costly. In this case, if the invention is useful enough, a company may proceed and pass on the added cost to consumers.

International economic tension has also been increasing as a result of differing policies. Many foreign individuals and companies believe that they are at a serious disadvantage in the United States with regard to patent ownership because of the logistical difficulty in establishing first-to-invent status. Further, failure of the United States to recognize patent ownership in other countries is in violation of the Paris Conventions on Industrial Properties, which requires all member nations to treat all patents equally. The conflict surrounding patents has prompted the World Intellectual Properties Organization (WIPO) to lobby for universality in patent laws. WIPO maintains that the first necessary step involves compelling the United States to reexamine its patent principle, taking into account the reality of a global economy. This push may indeed result in more global economic cooperation.

Summary

In his paper "Global Implications of Patent Law Variation," Koji Suzuki (1991) states that lack of consistency in the world's patent laws is a serious problem. In most of the world, patent ownership is given to the inventor that is first to file for a patent. However, the United States maintains a first-to-invent policy. In view of this, patent ownership can change depending on the country. Multiple patent ownership can result in economic problems; however, most striking is the international tension it causes. The fact that the United States does not recognize patent ownership in other countries, in violation of the Paris Convention on Industrial Properties, has prompted the World Intellectual Properties Organization (WIPO) to push the United States to review its existing patent law principles.

Part 3   Listening and Speaking

⟫ Dialogue: Installing Oracle Database Software

(As a newcomer, Henry was confused with a mass of complicated introduction documents about Oracle database when trying to install the Oracle software on his computer for the first time.)

Henry	Excuse me, Sophie and Mark. Do you know whether there are any quick ways for installing the Oracle software?[1]
Sophie	Well, Henry, I think you can use the Oracle Universal Installer. It can handle the complex internals and guide the beginning of DBA through the database installing process.
Mark	As Sophie says, Oracle Universal Installer is a GUI tool that enables you to *intuitively* view the Oracle software that is installed on your machine, install new Oracle software, and delete Oracle software that you no longer intend to use.
Henry	Like an installation engine, right?
Mark	Exactly! Before installation, Oracle Universal Installer checks the environment to see whether it meets the requirements for successful installation.[2] Early detection of problems with the system setup reduces the chances of users encountering problems during installation.
Henry	Then, what are the minimum system requirements?
Mark	The system requirements for Oracle Universal Installer broadly include Java Runtime Environment, Memory Requirements and Disk Space Requirements. Further more, you can check the Release *Notes* or installation guide for the products that you are installing for details.
Henry	How can I get the installer?
Sophie	You can install the Oracle software from the Oracle CD. If you don't have access to it, you can also download the Oracle Database software from Oracle Corporation's Website. It is free for non-commercial use, but you

	will need to purchase a license if you want to use it for purposes other than learning.
Henry	Are there any *tips* or notes when executing the installing program?
Sophie	To start the Oracle installer on a Windows environment, you simply put the Oracle Database CD in the CD-ROM drive and the autorun will start the installer.
Henry	Sounds very easy.
Mark	Yes. The Oracle Universal Installer will guide you through the installation process. It makes the installation process quite easy and you simply follow the prompts. [3]
Sophie	However, at any time while installing your product, click Help for information about the screens specific to your installation.
Henry	What kinds of help information are mostly provided?
Mark	Oracle Universal Installer provides two kinds of online help. One is Generic online help provided with every copy of Oracle Universal Installer. These topics *troubleshoot* the problems that every Oracle Universal Installer user encounters. Another is online help specific to a particular installation. These topics are often custom helps created by the product developers and describe the screens and dialog boxes specific to the product you are installing.[4]
Henry	I see! And really appreciate your great help.

Exercises

Work in pairs, and make up a similar conversation by replacing the scenarios with other material in the below.

Installing Sybase Adaptive Server Database

[1] Installing Sybase Adaptive Server database software on Linux is fairly straightforward if you already know a little about Linux, but it can be a little confusing for newcomers.

[2] There are two things you have to do before installing:
- Check that your current operating system configuration will allow Sybase Adaptive Server Database to install and run, since Sybase Adaptive Server Database on Linux is officially supported on a fairly specific set of distributions.
- Make sure that your system is configured properly, and that the correct set of software

packages is installed.

Do those by following the Sybase installation instructions in the Sybase Adaptive Server Database Installation Guide.

[3] When Performing the Installation and Configuration.

- Download the .tgz file from Sybase Website
- *Unpack* this file in a temporary directory.
- The installation can be run in GUI mode, or in console mode.Run the setup program—run "setup" to be GUI mode, or "setup –console" in text/terminal mode, and follow the prompts.
- If you want to use **ODBC** on the Linux, you'll need to use either the "Full" or the "Custom" installation type—otherwise the "Typical" installation is probably fine.
- Once the installation has completed you will be prompted to create an initial set of servers. These can be configured automatically using standard *defaults*, or manually by custom configuration.

[4] If you *run into* a problem, start by checking the installation and or server error logs. One or both of these files should have information about the reason why the installation failed. Take that information, and start with the Sybase Adaptive Server Database on Linux **FAQ**, which is a good way of getting help.

Words

intuitively[in'tjuitivli] *n*.直觉地，直观地	**troubleshoot**['trʌblʃu:t] *v*.检修
note[nəut] *n*.注解，注释	**unpack**['ʌn'pæk] *v*.打开取出
tip[tip] *n*.指点，提示	**default**[di'fɔ:lt] *n*.默认（值）

Phrases

run into	遇到，陷入

Abbreviations

ODBC	Open DataBase Connectivity	开放数据库互连
FAQ	Frequently Asked Question	常见问题解答

≫ Listening Comprehension: Data Mining

Listen to the passage and the following 3 questions based on it. After you hear a question, there will be a break of 10 seconds. During the break, you will decide which one is the best answer among the four choices marked (A), (B), (C) and (D).

Questions

1. Where does data mining derives its name from?

2. What is the ultimate purpose of data mining?

3. What measures did the retailer take after analyzing local buying patterns using data mining?

Choices

1. (A) Data analyzing   (B) Mountain excavating

    (C) Knowledge discovery   (D) Consumer focus

2. (A) Analyzing enormous sets of data   (B) Calculating an immense amount of value

    (C) Predicting behaviors and future trend   (D) Producing market research reports

3. (A) Move the beer display closer to the diaper display

    (B) Sell beer at a discount

    (C) Move the beer display away from the diaper display

    (D) Sell diapers at a discount

 Words

proactive[ˌprəʊˈæktiv] *adj.*主动的，先发制人的	**spot**[spɒt] *v.*发现，准确地定出…的位置
vein[vein] *n.*矿脉，特色，风格	**loyalty**[ˈlɔiəlti] *n.*忠诚，忠实
ore[ɔː(r)] *n.*矿石	**grocery**[ˈgrəʊsəri] *n.*杂货店
immense[iˈmens] *adj.*极大的，巨大的	**diaper**[ˈdaiəpə] *n.*尿布
probe[prəub] *v.*探查，查明	**genetics**[dʒiˈnetiks] *n.*遗传学

>> Dictation: Data Warehouse

This passage will be played THREE times. Listen carefully, and fill in the blanks with the words you have heard.

The term "Data Warehouse" was **coined** by W. H. Inmon, who is _____1_____ recognized as the "father of the data warehouse" in 1990, while IBM sometimes uses the term "information warehouse." However, the concept of data warehousing can _____2_____ to the late-1980s when IBM researchers Barry Devlin and Paul Murphy developed the "business data warehouse".

Based on analogies with _____3_____ warehouses, data warehouses were intended as large-scale _____4_____ /storage/*staging areas* for corporate data. Data could be retrieved from one _____5_____ point or data could be _____6_____ to "retail stores" or "data marts" which were tailored for ready _____7_____ by users.

Data warehouses are designed to _____8_____ reporting and analysis. An expanded _____9_____ for data warehousing includes business intelligence tools, tools to extract, _____10_____ , and load data into the *repository*, and tools to manage and retrieve *metadata*. In essence, it was intended to provide an _____11_____ model for the flow of data from _____12_____ systems to_____13_____ environments, with attempt to _____14_____ the various problems associated with this flow—mainly, the _____15_____ associated with it.

The term data warehouse also generally refers to the combination of many different_____16_____ across an entire _____17_____ . It contains a wide variety of data that _____18_____ a *coherent* picture of business conditions at a single point in time. Today, data warehouse has emerged as an important way for an organization to use data to come up with facts, _____19_____ or relationships that can help them make effective decisions or create effective strategies to _____20_____ their goals.

Words

coin[kɔin] *v.*杜撰，设计	**metadata**['metə'deitə] *n.*元数据
repository[ri'pɔzitəri] *n.*仓库	**coherent**[kəu'hiərənt] *adj.*一致的，连贯的

Phrases

staging area	临时数据交换区

Computer Network

Part 1   **Reading and Translating**

≫ **Section A   Ethernet**

Ethernet is a family of frame-based computer networking technologies for local area networks (LANs). The name comes from the physical concept of the **ether**. *It defines a number of wiring and signaling standards for the physical layer, through means of network access at the Media Access Control (**MAC**)/Data Link Layer, and a common addressing format.*

Ethernet is standardized as **IEEE** 802.3. *The combination of the twisted pair versions of Ethernet for connecting end systems to the network, along with the fiber optic versions for site **backbones**, is the most widespread wired LAN technology.* It has been in use from around 1980 to the present, largely replacing competing LAN standards such as token ring, **FDDI**, and **ARCNET**.

◆History

Ethernet was originally developed at Xerox PARC in 1973–1975. In

1975, Xerox filed a patent application listing Robert Metcalfe and David Boggs, plus Chuck Thacker and Butler Lampson, as inventors (U.S. Patent 4,063,220: Multipoint data communication system with collision detection). In 1976, after the system was deployed at PARC, Metcalfe and Boggs published a *seminal paper*.

The experimental Ethernet described in that paper ran at 3 Mbit/s, and had 8-bit destination and source address fields, so Ethernet addresses were not the global addresses they are today. By software convention, the 16 bits after the destination and source address fields were a packet type field, but, as the paper says, "different protocols use *disjoint* sets of *packet* types", so those were packet types within a given protocol, rather than the packet type in current Ethernet which specifies the protocol being used.

Metcalfe left Xerox in 1979 to promote the use of personal computers and local area networks (LANs), *forming* 3Com. He convinced **DEC**, Intel, and Xerox to work together to promote Ethernet as a standard, the so-called "**DIX**" standard, for "Digital/Intel/Xerox" ; it standardized the 10 megabits/second Ethernet, with 48-bit destination and source addresses and a global 16-bit type field. The standard was first published on September 30, 1980. It competed with two largely proprietary systems, token ring and ARCNET, which soon found themselves buried under a *tidal* wave of Ethernet products. In the process, 3Com became a major company.

Twisted-pair Ethernet systems have been developed since the mid-80s, beginning with StarLAN, but becoming widely known with 10BASE-T. *These systems replaced the **coaxial** cable on which early Ethernets were deployed with a system of **hubs** linked with unshielded twisted pair (**UTP**), ultimately replacing the **CSMA/CD** scheme in favor of a switched full **duplex** system offering higher performance.*

⊖ **General Description**

A 1990s Ethernet network interface card supports both coaxial-based using a 10BASE2 (**BNC** connector, left) and twisted pair-based 10BASE-T, using a RJ45 (8P8C modular connector, right).

Ethernet was originally based on the idea of computers communicating over a shared coaxial cable as a broadcast transmission medium. The methods used show some similarities to radio systems, although there are fundamental differences, such as the fact that it is much easier to detect collisions in a cable broadcast system than a radio broadcast. The common cable providing the communication channel was *likened* to the ether and it was from this reference that the name "Ethernet" was derived.

From this early and comparatively simple concept, Ethernet evolved into the complex networking technology that today underlies most LANs. The coaxial cable was replaced with point-to-point links connected by Ethernet hubs and/or *switches* to reduce installation costs, increase reliability, and enable point-to-point management and troubleshooting. StarLAN was the first step in the evolution of Ethernet from a coaxial cable *bus* to a hub-managed, twisted-pair network. The *advent* of twisted-pair wiring *dramatically* lowered installation costs relative to competing technologies, including the older Ethernet technologies.

Above the physical layer, Ethernet stations communicate by sending each other data packets, blocks of data that are individually sent and delivered. *As with* other IEEE 802 LANs, each Ethernet station is given a single 48-bit MAC address, which is used both to specify the destination and the source of each data packet. Network interface cards (**NIC**s) or chips normally do not accept packets addressed to other Ethernet stations. *Adapters generally come programmed with a globally unique address, but this can be overridden, either to avoid an address change when an adapter is replaced, or to use locally administered addresses.*

Despite the significant changes in Ethernet from a thick coaxial cable bus running at 10 Mbit/s to point-to-point links running at 1 Gbit/s and beyond, all generations of Ethernet (excluding early experimental versions) share the same frame formats (and hence the same interface for higher layers), and can be readily interconnected.

Due to the *ubiquity* of Ethernet, the ever-decreasing cost of the hardware needed to support it, and the reduced panel space needed by twisted pair Ethernet, most manufacturers now build the functionality of an Ethernet card directly into PC motherboards, *obviating* the need for installation of a separate network card.

 Words

Ethernet['i:θənet] *n.*以太网	**hub**[hʌb] *n.*网络集线器
ether['i:θə] *n.*以太	**duplex**['dju:pleks] *adj.*双工的
backbone['bækbəun] *n.*主干网，广域网中的一种高速链路	**liken**['laikən] *v.*把…比作
seminal['si:minl] *adj.*开创性的，有重大影响的	**switch**[switʃ] *n.*交换机
paper['peipə] *n.*论文，文章	**bus**[bʌs] *n.*总线
disjoint[dis'dʒɔint] *adj.*不相交的，没有交集的	**advent**['ædvənt] *n.*出现，到来
packet['pækit] *n.*包	**dramatically**[drəúmætikəli] *adv.*显著地，引人注目地
form[fɔ:m] *v.*建立，组成	**ubiquity**[ju:úbikwəti] *adv.*到处存在，普遍存在
tidal['taidl] *adj.*潮水般的	**obviate**['ɔbvieit] *v.*避免
coaxial[kəu'æksəl] *adj.*同轴的	

 Phrases

twisted pair	双绞线	**as with**	正如……一样
token ring	令牌环网		

Abbreviations

MAC	Media Access Control	介质访问控制
IEEE	Institute of Electrical and Electronics Engineers	美国电气和电子工程师协会
FDDI	Fiber Distributed Data Interface	光纤分布式数据接口
ARCNET	Attached Resource Computer NETwork	附加资源计算机网络
DEC	Digital Equipment Corporation	美国数字设备公司
DIX	Digital/Intel/Xerox	数字设备/英特尔/施乐公司
UTP	Unshielded Twisted Pair	非屏蔽双绞线
CSMA/CD	Carrier Sense Multiple Access/Collision Detection	载波侦听多路访问/冲突检测
BNC	Bayonet Neill-Concelman	尼尔-康塞曼插刀
NIC	Network Interface Card	网络接口卡

Complex Sentences

1. **Original:** The combination of the twisted pair versions of Ethernet for connecting end systems to the network, along with the fiber optic versions for site backbones, is the most widespread wired LAN technology.

   Translation: 以太网的用于将终端系统连接到网络的双绞线版本与用于场地主干网的光纤版本的组合，是最为广泛使用的有线局域网技术。

2. **Original:** These systems replaced the coaxial cable on which early Ethernets were deployed with a system of hubs linked with unshielded twisted pair (UTP), ultimately replacing the CSMA/CD scheme in favor of a switched full duplex system offering higher performance.

   Translation: 这些系统用以非屏蔽双绞线链接的网络集线器系统替代了基于同轴电缆所部署的早期以太网，并最终选择一个能提供更高性能的可交换全双工系统替代了载波侦听多路访问/冲突检测模式。

Exercises

I. **Read the following statements carefully, and decide whether they are true (T) or false (F) according to the text.**

   ___1. The name "Ethernet" was derived from the similar fundamentals between computers communicating over a shared coaxial cable and radio broadcast systems over the ether.

   ___2. Ethernet, token ring, FDDI, and ARCNET LAN are all LAN standards.

   ___3. Data packets are blocks of data that are individually sent and delivered, via which Ethernet stations communicate with each other.

   ___4. A specific protocol can be identified via the packet type it uses in current Ethernet.

   ___5. Now, most manufacturers build smaller network card with decreasing cost and the reduced panel space in PC.

II. Choose the best answer to each of the following questions.

1. Which of the following has been outdated among wired LAN technology?
   (A) The shielded twisted pair
   (B) The fiber optic
   (C) The coaxial cable
   (D) The unshielded twisted pair

2. Which of the following is WRONG about the comparison between the early version and the current version of Ethernet?
   (A) The original address fields in the experimental Ethernet have been changed in the current versions.
   (B) The destinstion and source adresses have been overleapt in the current Ethernet　versions.
   (C) A switched full duplex system can offer higher performance than the CSMA/CD scheme used by the early Ethernet.
   (D) The running speed of the current Ethernet is 100 times and beyond as fast as that of the early versions.

3. What does the phrase "frame-based" mean appearing at the beginning of the article?
   (A) All generations of Ethernet conform to the multi-layers structure of the OSI reference model.
   (B) All generations of Ethernet are built on improving infrastructure from coaxial cable to twisted pair.
   (C) All generations of Ethernet base on the same architecture formats and the same interfaces for higher layers.
   (D) All generations of Ethernet base on the same physical layer as the underlying frame.

III. Translating.

1. Ethernet defines a number of wiring and signaling standards for the physical layer, through means of network access at the Media Access Control (MAC)/Data Link Layer, and a common addressing format.

2. Adapters generally come programmed with a globally unique address, but this can be overridden, either to avoid an address change when an adapter is replaced, or to use locally administered addresses.

>> Section B　Wi-Fi

Wi-Fi is the trade name for the popular wireless technology used in home networks, mobile phones, video games and more. In particular, it covers the various IEEE 802.11 technologies (including 802.11n, 802.11b, 802.11g, and 802.11a). Wi-Fi technologies are supported by nearly every modern personal computer operating system and most advanced game consoles, printers, and other *peripherals*.

⬀Purpose

The purpose of Wi-Fi is to hide complexity by enabling wireless access to applications and data, media and streams. The main aims of Wi-Fi are the following:

- make access to information easier
- ensure compatibility and co-existence of devices
- eliminate cabling and wiring
- eliminate switches, adapters, plugs, pins and connectors

Uses

A Wi-Fi enabled device such as a PC, game console, mobile phone, MP3 player or PDA can connect to the Internet within range of a wireless network connected to the Internet. *The coverage of one or more interconnected access points — called a hotspot — can comprise an area as small as a single room with wireless-***opaque** *walls or as large as many square miles covered by overlapping access points.* Wi-Fi technology has served to set up *mesh* networks, for example, in London. Both architectures can operate in community networks.

In addition to restricted use in homes and offices, Wi-Fi can make access publicly available at Wi-Fi hotspots provided either free of charge or for *subscribers* of various providers. Organizations and businesses such as airports, hotels and restaurants often provide free hotspots to attract or assist clients. Enthusiasts or authorities who wish to provide services or even to promote business in a given area sometimes provide free Wi-Fi access. Metropolitan-wide Wi-Fi (**Muni-Fi**) already has more than 300 projects in process. There were 879 Wi-Fi based Wireless Internet Service Providers in the Czech Republic as of May 2008.

Wi-Fi also allows connectivity in peer-to-peer (wireless *ad-hoc* network) mode, which enables devices to connect directly with each other. This connectivity mode can prove useful in consumer electronics and gaming applications.

When wireless networking technology first entered the market many problems *ensued* for consumers who could not rely on products from different vendors working together. The Wi-Fi Alliance began as a community to solve this issue — aiming to address the needs of the end-user and to allow the technology to mature. The Alliance created the *branding* Wi-Fi *Certified* to *reassure* consumers that products will interoperate with other products displaying the same branding.

Many consumer devices use Wi-Fi. *Amongst others,* personal computers can network to each other and connect to the Internet, mobile computers can connect to the Internet from any Wi-Fi hotspot, and digital cameras can transfer images wirelessly.

Routers which incorporate a *DSL-modem* or a cable-modem and a Wi-Fi access point, often set up in homes and other *premises*, provide Internet-access and internetworking to all devices connected (wirelessly or by cable) to them. One can also connect Wi-Fi devices in ad-hoc mode for client-to-client connections without a router.

As of 2007 Wi-Fi technology had spread widely within business and industrial sites. *In business environments, just like other environments, increasing the number of Wi-Fi access-points provides redundancy, support for fast **roaming** and increased overall network- capacity by using more channels or by defining smaller cells.* Wi-Fi enables wireless voice- applications (**VoWLAN** or **WVOIP**). *Over the years, Wi-Fi implementations have moved toward "thin" access-points, with more of the network intelligence housed in a centralized network appliance, **relegating** individual access-points to the role of mere "**dumb**" radios.* Outdoor applications may utilize true mesh **topologies**. As of 2007 Wi-Fi **installations** can provide a secure computer networking **gateway**, **firewall**, **DHCP** server, **intrusion** detection system, and other functions.

Advantages and Challenges

Wi-Fi allows LANs (Local Area Networks) to be deployed without cabling for client devices, typically reducing the costs of network deployment and expansion. Spaces where cables cannot be run, such as outdoor areas and historical buildings, can **host** wireless LANs.

In 2008, wireless network adapters are built into most modern laptops. The price of chipsets for Wi-Fi continues to drop, making it an economical networking option included in ever more devices. Wi-Fi has become widespread in corporate infrastructures.

Different competitive brands of access points and client network interfaces are inter-operable at a basic level of service. Products **designated** as "Wi-Fi Certified" by the Wi-Fi Alliance are backwards compatible. Wi-Fi is a global set of standards. Unlike mobile telephones, any standard Wi-Fi device will work anywhere in the world.

Wi-Fi is widely available in more than 220,000 public hotspots and tens of millions of homes and corporate and university campuses worldwide. **WPA** is not easily **cracked** if strong passwords are used and WPA2 encryption has no known weaknesses. New protocols for Quality of Service (**WMM**) make Wi-Fi more suitable for **latency**-sensitive applications (such as voice and video), and power saving mechanisms (WMM Power Save) improve battery operation.

Wi-Fi networks have limited range. A typical Wi-Fi home router using 802.11b or 802.11g with a **stock** antenna might have a range of 32 m (120 ft) indoors and 95 m (300 ft) outdoors. Range also varies with frequency band. Wi-Fi in the 2.4 GHz frequency **block** has slightly better range than Wi-Fi in the 5 GHz frequency block. Outdoor range with improved (directional) antennas can be several kilometers or more with *line-of-sight*.

Wi-Fi performance decreases roughly **quadratically** as the range increases at constant radiation levels.

Threats to Security

The most common wireless encryption standard, Wired Equivalent Privacy or **WEP**, has been shown to be easily breakable even when correctly configured. Wi-Fi Protected Access (WPA and WPA2), which began shipping in 2003, aims to solve this problem and is now available on most products. Wi-Fi Access Points typically default to an "open" (encryption-free) mode. *Novice* users benefit from a zero-configuration device that works out of the box, but this default is without any wireless security enabled, providing open wireless access to their LAN. To turn security on requires the user to configure the device, usually via a software graphical user interface (GUI). Wi-Fi networks that are open (unencrypted) can be monitored and used to read and copy data (including personal information) transmitted over the network, unless another security method is used to secure the data, such as a **VPN** or a secure Web page.

History

Wi-Fi uses both single *carrier* direct-sequence spread *spectrum* radio technology (part of the larger family of spread spectrum systems) and multi-carrier **OFDM** (*Orthogonal* Frequency Division *Multiplexing*) radio technology. The regulations for unlicensed spread spectrum enabled the development of Wi-Fi, its *onetime* competitor Home**RF**, *Bluetooth*, and many other products such as some types of *cordless* telephones.

Unlicensed spread spectrum was first made available in the US by the Federal Communications Commission in 1985 and these **FCC** regulations were later copied with some changes in many other countries enabling use of this technology in all major countries. The FCC action was proposed by Michael Marcus of the FCC in 1980 and the subsequent regulatory action took 5 more years. It was part of a broader proposal to allow civil use of spread spectrum technology and was opposed at the time by main stream equipment manufacturers and many radio system operators.

The *precursor* to Wi-Fi was invented in 1991 by NCR Corporation/AT&T (later Lucent & Agere Systems) in Nieuwegein, the Netherlands. It was initially intended for cashier systems; the first wireless products were brought on the market under the name WaveLAN with speeds of 1 Mbit/s to 2 Mbit/s. Vic Hayes, who held the chair of IEEE 802.11 for 10 years and has been named the "father of Wi-Fi," was involved in designing standards such as IEEE 802.11b, and 802.11a.

Words

peripheral[pəˈrifərəl] *n.*外围设备	**firewall**[ˈfaiəwɔːl] *n.*防火墙
opaque[əuˈpeik] *adj.*不透射的，不传导的	**intrusion**[inˈtruːʒən] *n.*闯入，侵扰

mesh[meʃ] *n.*网状物

subscriber[sʌbsˈkraibə] *n.*订购者，用户

ad-hoc[ˈædˈhɔk] *adj.*特别的，特定的

ensue[inˈsjuː] *v.*相继发生

branding[brændiŋ] *n.*商标，品牌

certify[ˈsəːtifai] *v.*证明，确认

reassure[riːəˈʃuə] *v.*使…恢复信心，打消…的疑虑

router[ˈruːtə] *n.*路由器

modem[ˈməudəm] *n.*调制解调器

premise[ˈpremis] *n.*建筑物

roam[rəum] *v.*漫游

relegate[ˈreligeit] *v.*把…降级，把…归类

dumb[dʌm] *adj.*哑的，不智能的，被动的

topology[təˈpɔlədʒi] *n.*拓扑

installation[ˌinstəˈleiʃən] *n.*装置，设备

gateway[ˈgeitwei] *n.*网关

host[həust] *v.*做主人招待，托管

designate[ˈdezigneit] *v.*选定，指派

crack[kræk] *v.*解开（秘密等）

latency[ˈleitənsi] *n.*延迟，等待时间

stock[stɔk] *n.*坐（式）

block[blɔk] *n.*区域

quadratically[kwəˈdrætikəli] *adv.*二次地，平方地

novice[ˈnɔvis] *n.*新手，初学者

carrier[ˈkæriə] *n.*载波（信号）

spectrum[ˈspektrəm] *n.*光谱

orthogonal[ɔːˈθɔgənl] *adj.*正交的

multiplex[ˈmʌltipleks] *adj.*多路传输的

onetime[ˈwʌntaim] *adj.*从前的，以前的

bluetooth[ˈbluːtuːθ] *n.*蓝牙

cordless[ˈkɔːdlis] *adj.*无绳的

precursor[pri(ː)ˈkəːsə] *n.*先驱，前任

Phrases

amongst others	其中	line-of-sight	视线，瞄准线

Abbreviations

Muni-Fi	Metropolitan-wide Wi-Fi	城域Wi-Fi网
DSL	Digital Subscriber Line	数字用户线路
VoWLAN	Voice over WLAN(Wireless Local Area Networks)	基于无线局域网络的IP语音通信
WVOIP	Wireless VoIP(Voice over IP)	无线VoIP
DHCP	Dynamic Host Configuration Protocol	动态主机分配协议
WPA	Wi-Fi Protected Access	Wi-Fi保护访问
WMM	Wireless MultiMedia	无线多媒体
WEP	Wired Equivalent Privacy	有线对等保密
VPN	Virtual Private Network	虚拟专用网络
OFDM	Orthogonal Frequency Division Multiplexing	正交频分复用技术
RF	Radio Frequency	无线射频
FCC	Federal Communications Commission	美国联邦通讯管理委员会

 Complex Sentences

1. **Original:** In business environments, just like other environments, increasing the number of

Wi-Fi access-points provides redundancy, support for fast roaming and increased overall network-capacity by using more channels or by defining smaller cells.

Translation: 在商业环境中，正如其他环境一样，不断增加的Wi-Fi访问点数量通过使用更多的信道或定义更小的单元来提供冗余、对快速漫游的支持以及更大的总网络容量。

2. **Original:** Over the years, Wi-Fi implementations have moved toward "thin" access-points, with more of the network intelligence housed in a centralized network appliance, relegating individual access-points to the role of mere "dumb" radios.

Translation: 经过多年，Wi-Fi实现已经移向了"瘦"访问点，使用设在一个集中网络装置中的更多网络智能，将单独访问点降级到只是"哑的"无线电设备的角色。

Exercises

I. Read the following statements carefully, and decide whether they are true (T) or false (F) according to the text.

____1. A Wi-Fi enabled device can connect to the Internet when within range of any wireless network.

____2. Wi-Fi can make access publicly available at Wi-Fi hotspots via either free of charge or subscription.

____3. Wi-Fi enabled devices need a router in all modes for connections.

____4. Wi-Fi in the higher GHz frequency block has better range than in the lower GHz frequency block.

____5. You can use any standard Wi-Fi device anywhere in the world.

II. Choose the best answer to each of the following questions.

1. Which of the following is WRONG about Wi-Fi?

(A) It can be broadly divided into two main radio technologies.

(B) It was originated with the regulations for unlicensed spread spectrum.

(C) It was a overall proposal which allows civil use of spread spectrum technology.

(D) It has become an economical and widespread networking option since the cost of hardware continues to drop.

2. Which of the following is NOT the method or standard used to secure the Wi-Fi network?

(A) WEP   (B) WPA

(C) VPN   (D) WMM

3. Which of the following is the problem most urgently to be solved when wireless networking technology first entered the market?

(A) Implementing the interoperation of products displaying the same branding.

(B) Making Wi-Fi networks free of threats to security.

(C) Providing free Wi-Fi access publicly available at Wi-Fi hotspots to attract users.

(D) Competing with and exceeding its rival such as HomeRF and Bluetooth.

III. Translating.

1. The coverage of one or more interconnected access points — called a hotspot — can comprise an area as small as a single room with wireless-opaque walls or as large as many square miles covered by overlapping access points.

2. Routers which incorporate a DSL-modem or a cable-modem and a Wi-Fi access point, often set up in homes and other premises, provide Internet-access and internetworking to all devices connected (wirelessly or by cable) to them.

Part 2   Simulated Writing: Instructions

Introduction

One of the most common and one of the most important uses of technical writing is instructions—those step-by-step explanations of how to do things: assemble something, operate something, repair something, or do routine maintenance on something. What follows in this part will show you what professionals consider the best techniques. You can in turn use these considerations to plan your own instructions.

Ultimately, however, good instruction writing not only requires these techniques but also:

- Clear, simple and specific writing
- A thorough understanding of the procedure in all its technical details
- Your ability to put yourself in the place of the reader, the person trying to use your instructions
- Your ability to visualize the procedure in great detail and to capture that awareness on paper
- Finally, your willingness to go that extra distance and test your instructions on the kind of person you wrote them for

Some Preliminaries

At the beginning of a project to write instructions, it's important to determine the structure or characteristics of the particular procedure you are going to write about.

Audience and situation. Early in the process, define the audience and situation of your instructions. Remember that defining an audience means defining its level of familiarity with the topic as well as other such details. Write for your audience and use a level of detail that is suitable to their skill level.

Number of tasks. An important consideration is how many tasks there are in the procedure you are writing instructions for. Let's use the term procedure to refer to the whole set of activities your instructions are intended to discuss, while a task is a semi-independent group of actions within the procedure.

Best approach to the step-by-step discussion. Another consideration, which maybe you can't determine early on, is how to focus your instructions. For most instructions, you can focus on tasks, or you can focus on tools (or features of tools). You can use a task approach or a tools approach to instructions.

Groupings of tasks. Listing tasks may not be all that you need to do. There may be so many tasks that you must group them so that readers can find individual ones more easily.

Common Sections

The following is a review of the sections you'll commonly find in instructions.

Introduction. Plan the introduction to your instructions carefully. Make sure it does any of the following things (but not necessarily in this order) that apply to your particular instructions:

- Indicate the specific tasks or procedure to be explained as well as the scope of coverage (what won't be covered).
- Indicate what the audience needs in terms of knowledge and background to understand the instructions.
- Give a general idea of the procedure and what it accomplishes.
- Indicate the conditions when these instructions should (or should not) be used.
- Give an overview of the contents of the instructions.

General warning, caution, danger notices. Instructions often must alert readers to the possibility of ruining their equipment, screwing up the procedure, and hurting themselves. Also, instructions must often emphasize key points or exceptions. For these situations, you use special notices—note, warning, caution, and danger notices.

Technical background or theory. At the beginning of certain kinds of instructions (after the introduction, of course), you may need a discussion of background related to the procedure. For certain instructions, this background is critical—otherwise, the steps in the procedure make no sense.

Equipment and supplies. Notice that most instructions include a list of the things you need to gather before you start the procedure. This includes equipment, the tools you use in the procedure and supplies, the things that are consumed in the procedure.

Discussion of the steps. When you get to the actual writing of the steps, there are several things to keep in mind.

(1) **Structure and format**. Normally, we imagine a set of instructions as being formatted as vertical numbered lists. And most are in fact. Normally, you format your actual step-by-step instructions this way. There are some variations, however, as well as some other considerations:

- **Fixed-order steps** are steps that must be performed in the order presented. These are numbered lists (usually, vertical numbered lists).
- **Variable-order steps** are steps that can be performed in practically any order. With this type, the bulleted list is the appropriate format.
- **Alternate steps** are those in which two or more ways to accomplish the same thing are presented. Alternate steps are also used when various conditions might exist. Use bulleted lists with this type, with OR inserted between the alternatives, or the lead-in indicating that alternatives are about to be presented.
- **Nested steps**. In some cases, individual steps within a procedure can be rather complex in their own right and need to be broken down into substeps. In this case, you indent further and sequence the substeps as a, b, c, and so on.
- **"Stepless" instructions.** And finally there exist instructions that really cannot use numbered vertical list. Some situations must be so generalized or so variable that steps cannot be stated.

(2) **Supplementary discussion**. Often, it is not enough simply to tell readers to do this or to do that. They need additional explanatory information such as how the thing should look before and after the step; why they should care about doing this step; what mechanical principle is behind what they are doing; even more micro-level explanation of the step—discussion of the specific actions that make up the step.

(3) **Writing style**. Notice how "real-world" instructions are written—they use a lot of imperative (command, or direct-address) kinds of writing; they use a lot of "you." That's entirely appropriate. You want to get in your reader's face, get her or his full attention. For that reason, instruction-style sentences sound like these: "Now, press the Pause button on the front panel to stop the display temporarily" and "You should be careful not to ..."

Format in Instructions

Headings. In your instructions, make good use of headings. Normally, you'd want headings for any background section you might have, and the equipment and supplies section; a general heading for the actual instructions section; and subheadings for the individual tasks or phases within that section. Normally it is wise to split the instructions into separate sections using headings identifying the purpose of each action.

Lists. Similarly, instructions typically make heavy use of lists, particularly numbered vertical lists for the actual step-by-step explanations. Simple vertical lists or two-column lists are usually good for the equipment and supplies section. In-sentence lists are good whenever you give an overview of things to come.

Graphics. Sometimes, words simply cannot explain the step. Illustrations are often critical to readers' ability to visualize what they are supposed to do. Use appropriate illustrations, label and caption them well. This can save words and illuminate the words that remain.

Special notices. In instructions, you must alert readers to possibilities in which they may damage their equipment, waste supplies, injure themselves or others—even seriously or fatally. Companies have been sued for lack of these special notices, for poorly written special notices, or for special notices that were out of place. Split the special notices into chunks and put them at the start of the instructions to avoid they are buried in long paragraphs.

Number, abbreviations, and symbols. Instructions also use plenty of numbers, abbreviations, and symbols.

Other Useful Information

Watch out for problems such as the following and devote special effort to producing lucid and well-organized instructions:

- Focuses on generalized people rather than individuals.
- Prefix the instructions with a clear heading that summarizes the task.
- Remember that no heading "Introduction" is needed between the title and the first paragraph. Remember not to use first-level headings in this assignment; start with the second level.
- Use the various types of lists wherever appropriate. In particular, use a numbered list when the order is important. Use a bulleted list (like this one) when the order is not important.
- Avoid lists of more than about ten steps. Break down long lists into two or more subtasks.
- Start each instruction with an active verb that instructs the reader to do something.
- Use simple present tense.
- Use the class style and format for all headings, lists, special notices, and graphics.
- Finally, test the instructions on people who are genuinely typical of the target audience and redraft the instructions in the light of the feedback.

Sample

A Beginner's Guide to Eudora Lite 1.5.4 for Windows 95

This is a beginner's guide to using Eudora Lite 1.5.4 for Windows 95. It covers everything you need to know to start using Eudora to send and receive e-mail, including: (1) essential settings and program configuration; (2) the basic tasks involved in composing and sending e-mail, receiving and reading messages, as well as what to do with them when your done; and (3) how to find more information about using Eudora once you've learned the basics.

Eudora Lite is freeware (software which may be used at no cost) that offers a set of tools and basic e-mail utilities for managing electronic mail messages. The program employs standard Windows graphical interface elements, so that anyone familiar with the Windows environment should find Eudora easy to navigate. The application is configured to access a specific account on a POP server—a computer connected to a network or the Internet running software that sends, receives and stores e-mail. Eudora then acts as the communication agent between the remote server and the local computer—sending and receiving mail messages to and from the POP account. Finally, Eudora offers a set of utilities for manipulating and organizing messages stored on the local system.

To use this guide you need only be familiar with the basics of mouse operation and navigating standard Windows interface objects such as menus, windows, dialog boxes, and toolbars.

Eudora Setup

Before you can begin using Eudora, you must make sure you have the right equipment and configure Eudora.

Getting started. To use these instructions, you'll need the following:

1. An IBM PC-compatible computer running under Windows 95.

2. A copy of the Eudora Lite application installed and currently running on your machine.

3. Internet access with a POP3 e-mail account.

Note: If you are not sure what kind of account you have, contact your Internet Service Provider. POP simply refers to the type of software used on the remote server where your e-mail is received and stored. Most e-mail accounts today use POP3 software.

Before you can use Eudora for completing basic e-mail tasks, you must tell the program a few things about yourself and your account. If Eudora is not running, you should start it now.

Configuring the program. Eudora needs very little information before you can begin using it to send and receive e-mail. You need only:

1. Open the Tools menu and select Options as shown in Figure 1. Eudora opens a dialog box with several option categories.

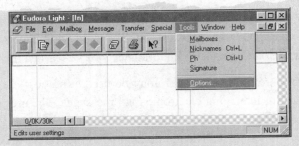

Figure 1   Tools Menu.

2. Choose the Getting Started icon in the Categories column.

3. Enter your POP Account in the space provided.

   Note: Your POP Account is your username followed by the "@" symbol, and the name of your e-mail server. For example, if your username is "jdoe" and your service provider is "the.mail.com", you should enter "jdoe@the.mail.com" in the POP Account space. This is your e-mail address.

4. Enter your actual name in the Real Name space. This name identifies you to your e-mail recipients.

5. Make sure the Winsock (Network, PPP, SLIP) button is checked.

When its settings resemble those shown in Figure 2, Eudora is ready to communicate with your e-mail server.

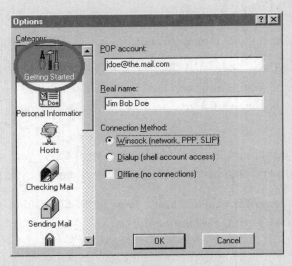

Figure 2   Basic Configuration Settings

Basic E-Mail Tasks

[The detailed content of each part is omitted from this sample.]

Getting More Information

[The detailed content of each part is omitted from this sample.]

Part 3   Listening and Speaking

⟫⟫ Dialogue: Setting up Wireless Network

(Sophie currently has her computer connected directly to a modem in her home. Now she wants to replace the current wired mode with a wireless network, and asks Henry and Mark for help.)

Henry	Take it easy, Sophie. It's easier to install than you think, I promise. There are only 4 steps in brief.
Sophie	Really? What's that?
Mark	The first step is to make sure that you have the equipment needed[1]. The list generally includes **broadband** Internet connection, a wireless router and a computer with built-in wireless network support or a wireless network adapter. [1]
Sophie	Let me see. OK, I think they are all ready now. And how about the next?
Mark	The next step is to connect your wireless router. After turning your modem off, unplug the network cable from the back of your computer, and plug it into the **port** labeled Internet, **WAN**, or **WLAN** on the back of your router. Meanwhile, the other end of the network cable should be plugged into your modem. [2]
Sophie	Well, how can I know whether they have been connected correctly?
Henry	It's easy. To check this, plug in and turn on your modem. Wait a few minutes to give it time to connect to the Internet, and then plug in and turn on your wireless router. After a minute, the Internet, WAN, or WLAN light on your wireless router should *light up*, indicating that it has successfully connected to your modem.
Sophie	Ok, so what else?

Henry	Then, configure your wireless router. You should temporarily connect your computer to one of the open network ports on your wireless router which isn't labeled Internet, WAN, or WLAN. Next, open Internet Explorer and type in the address and the password to configure your router.
Sophie	How can I get the address and password?
Henry	You can find them in the instructions included with your router.
Mark	Specially, in the process of configuration, Sophie, you should **mind** three things: your wireless network name which is known as the Service Set Identifier, wireless encryption or Wi-Fi Protected Access, and your administrative password which controls your wireless network. *By the way*, disconnect the network cable from your computer after the configuration is completed. [3]
Sophie	Yes, I've got it.
Henry	Now, we come into the final step, connecting your computers. Windows XP should show a wireless network icon with a **notification** that says it has found a wireless network. View available wireless networks list, choose your network and click Connect. Type the encryption key and then click Connect. Windows XP will show its progress as it connects to your network. After you're connected, you can now close the Wireless Network Connection window. [4] You're done for now.
Mark	Then you can use your laptop to surf the Web while you're sitting on your **couch** or in your *yard*. [5]
Sophie	Thanks a lot! I can't wait. Shall we start right now?
Henry & Mark	No problem.

Exercises

Work in pairs, and make up a similar conversation by replacing the scenarios with other material in the below.

Creating a LAN Connection between Two PCs

[1] Step 1: Prepare the equipment needed.
- Two computers
- A crossover (Cross **crimped** RJ45 / CAT5) cable

- A LAN (Ethernet) card in both computers

[2] Step 2: Connect 2 PCs with the cable.
Connect the cross crimped Ethernet cable in the LAN RJ45 ports of both the computers.

[3] Step 3: Configure the 2 PCs.
Configure the PC 1.
1. Change the names of PC1 and the workgroup.
  - Right-click the icon "My Computer" and click the item "Properties".
  - Select the "Computer Name" tab and click the "Change" button.
  - Change the computer and the workgroup names to whatever you want.
  - Note: The workgroup name should be the same in both computers.
  - Click OK.
2. Configure the TCP/IP connection settings.
  - Goto "Control Panel" >> "Network connections".
  - Right-click on your LAN connection and click "Properties". In the "General" tab, select "Internet Protocol (TCP/IP)". Click the "Properties" button..
  - Set the values:
    IP Address - 192.168.0.1
    Subnet Mask - 255.255.255.0
    Default Gateway - 192.168.0.2
  - Click OK and you are done for PC1.
Configure the PC 2.
The same as above, just one difference:
When configuring the TCP/IP connection settings of the PC2, set the values.
    IP Address - 192.168.0.2
    Subnet Mask - 255.255.255.0
    Default Gateway - 192.168.0.1

[4] If you have done the above part correctly, the computers will have detected each other. If they don't, just restart both the computers.

[5] You can access the other computer in "My Network Places" and play multi-player games using this connection.

 Words

broadband['brɔ:dbænd] *n*.宽带	**notification**[ˌnəutifiˈkeiʃən] *n*.通知，布告
port[pɔ:t] *n*.端口	**couch**[kautʃ] *n*.睡椅，沙发
mind[maind] *v*.注意，记住	**crimped**[krimpt] *adj*.起皱褶的，有波纹的

Phrases

light up	点燃，照亮	by the way	顺便提一句

Abbreviations

WAN	Wide Area Network	广域网
WLAN	Wireless Local Area Network	无线局域网络

>> Listening Comprehension: IPv6 - the Next Gteneration Internet Protocol

Listen to the passage and the following 3 questions based on it. After you hear a question, there will be a break of 10 seconds. During the break, you will decide which one is the best answer among the four choices marked (A), (B), (C) and (D).

Questions

1. What is the most direct and powerful driving force behind the development of IPv6?
2. How many unique IP addresses can be supported by IPv6 at most?
3. Why does IPv4 need to use the technique of network address translation?

Choices

1. (A) The explosive growth of networks
   (B) An increasing demand for IP addresses
   (C) An escalating demand for wireless devices
   (D) A mature technique of network address translation
2. (A) 32   (B) 2^32   (C) 128   (D) 2^128
3. (A) Providing interoperability between IPv4 and IPv6 hosts
   (B) Avoiding running out of the available address space
   (C) Extending the amount of available address space
   (D) Providing control information to route packets

Words

escalate['eskəleit] v.逐步增强

Phrases

packet-switched	包交换	run out of	用光，用完

>> Dictation: Router

This passage will be played THREE times. Listen carefully, and fill in the blanks with the words you have heard.

A router consists of a computer with at least two network _____1_____ cards supporting the IP _____2_____ , whose software and hardware are usually tailored to the tasks of _____3_____ and forwarding information. It may be used to connect two or more wired or _____4_____ IP networks, or an IP network to an Internet connection. _____5_____ , a router is a "Layer 3 _____6_____ " which operates at the network layer of the OSI reference _____7_____ .

The router receives _____8_____ from each interface via a network interface and _____9_____ the received packets to an appropriate _____10_____ network interface. By maintaining _____11_____ information in a piece of storage called the "routing table," it also has the ability to filter _____12_____ to ensure that _____13_____ packets are *discarded*.

The very first device that had fundamentally the same _____14_____ as a router does today, was the Interface Message Processor (**IMP**). IMPs were the devices that _____15_____ the **ARPANET**, the first packet _____16_____ network. The idea for a router (although they were called "gateways" at the time) _____17_____ came about through an international group of computer networking researchers called the International Network Working Group (**INWG**).

Many engineers believe that the use of a router provides better _____18_____ against *hacking* than a software _____19_____ , because no computer IP address are _____20_____ exposed to the Internet.

📖 Words

discard[disˈkɑːd] *v.*丢弃，抛弃	**hack**[hæk] *v.*黑客入侵

📖 Abbreviations

IMP	Interface Message Processor	接口报文处理器
ARPANET	Advanced Research Projects Agency Network	（美国）高级研究计划署网络
INWG	International Network Working Group	国际互联网工作组

The Internet

Part 1   Reading and Translating
             Section A   Grid Computing
             Section B   Web 2.0
Part 2   Simulated Writing: Proposal
Part 3   Listening and Speaking
             Dialogue: Enhancing Your Computer Security
             Listening Comprehension: History of Google
             Dictation: How Web Search Engines Work

Part 1   Reading and Translating

≫ Section A   Grid Computing

*Grid computing is a form of distributed computing whereby a "super and virtual computer" is composed of a **cluster** of networked, loosely-coupled computers, acting in concert to perform very large tasks.* This technology has been applied to computationally-intensive scientific, mathematical, and academic problems through volunteer computing, and it is used in commercial enterprises for such diverse applications as drug discovery, economic forecasting, *seismic* analysis, and *back-office* data processing in support of e-commerce and Web services.

What distinguishes grid computing from typical cluster computing systems is that grids tend to be more loosely coupled, heterogeneous, and geographically dispersed. Also, while a computing grid may be dedicated to a specialized application, it is often constructed with the aid of *general purpose* grid software libraries and *middleware*.

Grids can be categorized with a three stage model of departmental grids, enterprise grids and global grids. These correspond to a firm initially utilizing resources within a single group i.e. an engineering department connecting desktop machines, clusters and equipment. This progresses to enterprise grids where non-technical staff's computing resources can be used for *cycle-stealing* and storage. A global grid is a connection of enterprise and departmental grids which can be used in a commercial or collaborative manner.

⊙ Grids versus Conventional Supercomputers

"Distributed" or "grid" computing in general is a special type of parallel computing which relies on complete computers (with onboard CPU, storage, power supply, network interface, etc.) connected to a network (private, public or the Internet) by a conventional network interface, such as Ethernet. This is in contrast to the traditional *notion* of a supercomputer, which has many processors connected by a local high-speed computer bus.

The primary advantage of distributed computing is that each node can be purchased as commodity hardware, which when combined can produce similar computing resources to a multiprocessor supercomputer, but at a lower cost. This is due to the economies of scale of producing commodity hardware, compared to the lower efficiency of designing and constructing a small number of *custom* supercomputers. The primary performance disadvantage is that the various processors and local storage areas do not have high-speed connections. This arrangement is thus well-suited to applications in which multiple parallel computations can take place independently, without the need to communicate intermediate results between processors.

The high-end *scalability* of geographically dispersed grids is generally favorable, due to the low need for connectivity between nodes relative to the capacity of the public Internet.

There are also some differences in programming and deployment. It can be costly and difficult to write programs so that they can be run in the environment of a supercomputer, which may have a custom operating system, or require the program to address concurrency issues. If a problem can be adequately parallelized, a "thin" layer of "grid" infrastructure can allow conventional, standalone programs to run on multiple machines (but each given a different part of the same problem). This makes it possible to write and debug on a single conventional machine, and eliminates complications due to multiple instances of the same program running in the same shared memory and storage space at the same time.

⊙ Design Considerations and Variations

One feature of distributed grids is that they can be formed from computing resources belonging to multiple individuals or organizations (known as multiple administrative domains). This can facilitate commercial transactions, as in utility computing, or make it easier to assemble volunteer computing networks.

One disadvantage of this feature is that the computers which are actually performing the calculations might not be entirely trustworthy. The designers of the system must thus introduce measures to prevent **malfunctions** or **malicious** participants from producing false, misleading, or erroneous results, and from using the system as an attack **vector.** This often involves assigning work randomly to different nodes (presumably with different owners) and checking that at least two different nodes report the same answer for a given work unit. Discrepancies would identify malfunctioning and malicious nodes.

Due to the lack of central control over the hardware, there is no way to guarantee that nodes will not *drop out of* the network at random times. Some nodes (like laptops or *dialup* Internet customers) may also be available for computation but not network communications for unpredictable periods. These variations can be **accommodated** by assigning large work units (thus reducing the need for continuous network connectivity) and reassigning work units when a given node fails to report its results as expected.

The impacts of trust and availability on performance and development difficulty can influence the choice of whether to deploy onto a dedicated computer cluster, to idle machines internal to the developing organization, or to an open external network of volunteers or contractors.

In many cases, the participating nodes must trust the central system not to abuse the access that is being granted, by interfering with the operation of other programs, **mangling** stored information, transmitting private data, or creating new security holes. Other systems employ measures to reduce the amount of trust "client" nodes that must place in the central system such as placing applications in virtual machines.

Public systems or those crossing administrative domains (including different departments in the same organization) often result in the need to run on heterogeneous systems, using different operating systems and hardware architectures. With many languages, there is a tradeoff between investment in software development and the number of platforms that can be supported (and thus the size of the resulting network). Cross-platform languages can reduce the need to make this tradeoff, though potentially at the expense of high performance on any given node (due to run-time interpretation or lack of optimization for the particular platform).

Various middleware projects have created generic infrastructure, to allow diverse scientific and commercial projects to harness a particular associated grid, or for the purpose of setting up new grids. **BOINC** is a common one for academic projects seeking public volunteers.

In fact, the middleware can be seen as a layer between the hardware and the software. On top of the middleware, a number of technical areas have to be considered, and these may or may not be middleware independent. Example areas include **SLA** management, Trust and Security, **VO** management, License Management, ***Portals*** and Data Management. *These technical areas may be* *taken care of in a commercial solution, though the* cutting edge *of each area is often found within specific research projects examining the field.*

CPU Scavenging

CPU-*scavenging*, cycle-scavenging, cycle stealing, or shared computing creates a "grid" from the unused resources in a network of participants (whether worldwide or internal to an organization). Typically this technique uses desktop computer instruction cycles that would otherwise be wasted at night, during lunch, or even in the scattered seconds throughout the day when the computer is waiting for user input or slow devices.

Volunteer computing projects use the CPU scavenging model almost exclusively.

In practice, participating computers also donate some supporting amount of disk storage space, RAM, and network bandwidth, in addition to raw CPU power. Since nodes are *apt* to go "offline" from time to time, as their owners use their resources for their primary purpose, this model must be designed to handle such *contingencies*.

History

The term grid computing originated in the early 1990s as a *metaphor* for making computer power as easy to access as an electric power grid in Ian Foster and Carl Kesselmans seminal work, "The Grid: Blueprint for a new computing infrastructure".

CPU scavenging and volunteer computing were popularized beginning in 1997 by distributed.net and later in 1999 by SETI@home to harness the power of networked PCs worldwide, in order to solve CPU-intensive research problems.

The ideas of the grid (including those from distributed computing, object oriented programming, Web services and others) were brought together by Ian Foster, Carl Kesselman and Steve Tuecke, widely regarded as the "fathers of the grid." They led the effort to create the Globus *Toolkit* incorporating not just computation management but also storage management, security provisioning, data movement, monitoring and a toolkit for developing additional services based on the same infrastructure including agreement negotiation, notification mechanisms, *trigger* services and information *aggregation*. *While the Globus Toolkit remains the* ***defacto*** *standard for building grid solutions, a number of other tools have been built that* ***answer*** *some subset of services needed to create an enterprise or global grid.*

During 2007 the term cloud computing came into popularity, which is conceptually similar to the ***canonical*** Foster definition of grid computing (in terms of computing resources being consumed as electricity is from the power grid). Indeed grid computing is often (but not always) associated with the delivery of cloud computing systems.

Words

grid[grid] *n.*网格	**portal**['pɔ:təl] *n.*门户
cluster['klʌstə] *n.*集群	**scavenge**['skævindʒ] *v.*提取有用之物
seismic['saizmik] *adj.*地震的	**apt**[æpt] *adj.*易于…的
middleware['midlweə] *n.*中间件	**contingency**[kən'tindʒənsi] *n.*意外事故，偶然，可能性
notion['nəuʃən] *n.*概念，看法	**metaphor**['metəfə] *n.*隐喻，暗喻
custom['kʌstəm] *adj.*定做的，定制的	**toolkit**[tu:lkit] *n.*工具包，工具箱
scalability[ˌskeilə'biliti] *n.*可括缩性	**trigger**['trigə] *n.*触发
malfunction[mæl'fʌŋkʃən] *n.*故障	**aggregation**[ægri'geiʃən] *n.*聚合
malicious[mə'liʃəs] *adj.*恶意的	**defacto**[di:'fæktəu] *adj.*事实上的，实际上的
vector['vektə] *n.*力量，动力	**answer**['ɑ:nsə] *v.*满足，适用
accommodate[ə'kɔmədeit] *v.*和解，调和	**canonical**[kə'nɔnikəl] *adj.*规范的
mangle['mæŋgl] *v.*破坏，毁损	

Phrases

in concert	一致，一齐	**drop out of**	退出
back-office	后台	**dial-up**	拨号（上网）
general purpose	多功能的，多用途的	**cutting edge**	尖端
cycle-stealing	周期侵占		

Abbreviations

BOINC	Berkeley Open Infrastructure for Network Computing	伯克利开放式网络计算平台
SLA	Service-Level Agreement	服务等级协议
VO	Virtual Organization	虚拟组织

Complex Sentences

1. **Original:** The impacts of trust and availability on performance and development difficulty can influence the choice of whether to deploy onto a dedicated computer cluster, to idle machines internal to the developing organization, or to an open external network of volunteers or contractors.

   Translation: 性能和开发难度上的信任度和可用性能够影响是否选择部署在一个专用计算机集群上，或在开发组织内部的空闲机器上，还是选择部署在志愿者或承包人的一个开放的外部网络中。

2. **Original:** These technical areas may be taken care of in a commercial solution, though the cutting edge of each area is often found within specific research projects examining the field.

Translation: 这些技术领域可能在商业解决方案中得到应用，尽管每个领域的尖端通常出现在研究该领域的特定研究项目内。

Exercises

I. Read the following statements carefully, and decide whether they are true (T) or false (F) according to the text.

___1. The term grid computing derived from the reference to the electric power grid.

___2. Grids evolve through four stages: departmental grids, enterprise grids, global grids and cloud computing.

___3. Generic middleware enables diverse projects of different domains to harness a particular associated grid.

___4. Cross-platform languages can reduce the need to make the tradeoff between investment in software development and the number of platforms that can be supported.

___5. The middleware is dependent on the technical areas which have to be considered on top of it.

II. Choose the best answer to each of the following questions.

1. Which of the following is WRONG about the advantages of girds?

   (A) Loosely coupled   (B) High-speed connections

   (C) Heterogeneous   (D) High-end scalability

2. Which of the following is NOT mentioned in this article about the comparison between grids and conventional supercomputers?

   (A) Physical composing   (B) Economic cost

   (C) Reliability   (D) Performance

3. Which of the following is RIGHT about the working of grids?

   (A) Nodes will not drop out of the network at random once grids have assigned tasks to them.

   (B) In the CPU scavenging model, participating computers only donate raw CPU power.

   (C) To prevent malfunctions or malicious participants, grids assign appointed work to reliable nodes.

   (D) Grids can place applications in virtual machines to reduce the amount of trust "client" nodes that must place in the central system.

III. Translating.

1. Grid computing is a form of distributed computing whereby a "super and virtual computer" is composed of a cluster of networked, loosely-coupled computers, acting in concert to perform very large tasks.

2. While the Globus Toolkit remains the defacto standard for building grid solutions, a number

of other tools have been built that answer some subset of services needed to create an enterprise or global grid.

⫸ Section B　Web 2.0

Web 2.0 is a term describing changing trends in the use of World Wide Web technology and Web design that aim to enhance creativity, information sharing, and collaboration among users. These concepts have led to the development and evolution of Web-based communities and hosted services, such as social-networking sites, video sharing sites, *wikis*, blogs, and *folksonomies*. The term became notable after the first O'Reilly Media Web 2.0 conference in 2004. Although the term suggests a new version of the World Wide Web, it does not refer to an update to any technical specifications, but to changes in the ways software developers and end-users utilize the Web. *According to Tim O'Reilly, Web 2.0 is the business revolution in the computer industry caused by the move to the Internet as platform, and an attempt to understand the rules for success on that new platform.*

➲ Definition

Web 2.0 has numerous definitions. Basically, the term encapsulates the idea of the *proliferation* of interconnectivity and social interactions on the Web. Tim O'Reilly regards Web 2.0 as business *embracing* the Web as a platform and using its strengths (global audiences, for example). O'Reilly considers that Eric Schmidt's *abridged* slogan, don't fight the Internet, *encompasses* the essence of Web 2.0 — building applications and services around the unique features of the Internet, as opposed to building applications and expecting the Internet to suit as a platform (effectively "fighting the Internet").

In the opening talk of the first Web 2.0 conference, O'Reilly and John Battelle summarized what they saw as the themes of Web 2.0. *They argued that the Web had become a platform, with software above the level of a single device, **leveraging** the power of the "Long Tail", and with data as a driving force.* According to O'Reilly and Battelle, an architecture of participation where users can contribute Website content creates network effects. Web 2.0 technologies tend to foster innovation in the assembly of systems and sites composed by pulling together features from distributed, independent developers. (This could be seen as a kind of "open source" or possible "*Agile*" development process, consistent with an end to the traditional software adoption cycle, *typified* by the so-called "perpetual beta".)

Web 2.0 technology encourages lightweight business models enabled by *syndication* of content and of service and by *ease of picking-up* by early adopters.

O'Reilly provided examples of companies or products that *embody* these principles in his description of his four levels in the hierarchy of Web 2.0 sites:

- Level-3 applications, the most "Web 2.0"-oriented, only exist on the Internet, deriving their effectiveness from the inter-human connections and from the network effects that Web 2.0 makes possible, and growing in effectiveness *in proportion as* people make more use of them. O'Reilly gave eBay, Craigslist, Wikipedia, del.icio.us, Skype, dodgeball, and AdSense as examples.
- Level-2 applications can operate offline but gain advantages from going online. O'Reilly cited Flickr, which benefits from its shared photo-database and from its community-generated tag database.
- Level-1 applications operate offline but gain features online. O'Reilly pointed to Writely (now Google Docs & Spreadsheets) and iTunes (because of its music-store portion).
- Level-0 applications work as well offline as online. O'Reilly gave the examples of MapQuest, Yahoo! Local, and Google Maps (mapping-applications using contributions from users to advantage could rank as "level 2").

Non-Web applications like email, instant-messaging clients, and the telephone fall outside the above hierarchy.

In *alluding* to the version-numbers that commonly designate software upgrades, the phrase "Web 2.0" hints at an improved form of the World Wide Web. Technologies such as Weblogs (blogs), wikis, *podcasts*, **RSS** *feeds* (and other forms of many-to-many publishing), social software, and Web application programming interfaces (APIs) provide enhancements over read-only Websites. Stephen Fry, who writes a *column* about technology in the British Guardian newspaper, describes Web 2.0 as: an idea in people's heads rather than a reality. It's actually an idea that the reciprocity between the user and the provider is what's emphasized. In other words, genuine interactivity, if you like, simply because people can upload as well as download.

The idea of "Web 2.0" can also relate to a transition of some Websites from isolated information *silos* to interlinked computing platforms that function like locally-available software in the *perception* of the user. Web 2.0 also includes a social element where users generate and distribute content, often with freedom to share and re-use. This can result in a rise in the economic value of the Web to businesses, as users can perform more activities online.

Characteristics

Web 2.0 Websites allow users to do more than just retrieve information. They can build on the interactive facilities of "Web 1.0" to provide "network as platform" computing, allowing users to run software-applications entirely through a browser. Users can own the data on a Web 2.0 site and exercise control over that data. These sites may have an "architecture of participation" that encourages users to add value to the application as they use it. This stands in contrast to very old traditional Websites, the sort which limited visitors to viewing and whose content only the site's owner could modify. Web 2.0 sites often feature a rich, user-friendly interface based on **Ajax**, openlaszlo, Flex or similar rich media.

The concept of Web-as-participation-platform captures many of these characteristics. Bart Decrem, a founder and former CEO of Flock, calls Web 2.0 the "*participatory* Web" and regards the Web-as-information-source as Web 1.0.

The impossibility of excluding group-members who don't contribute to the provision of goods from sharing profits gives rise to the possibility that rational members will prefer to **withhold** *their contribution of effort and free-ride on the contribution of others.*

According to Best, the characteristics of Web 2.0 are: rich user experience, user participation, dynamic content, metadata, Web standards and scalability. Further characteristics, such as openness, freedom and collective intelligence *by way of* user participation, can also be viewed as essential attributes of Web 2.0.

⮕ Technology Overview

The sometimes complex and continually evolving technology infrastructure of Web 2.0 includes server-software, content-syndication, messaging-protocols, standards-oriented browsers with plug-ins and extensions, and various client-applications. The differing, yet complementary approaches of such elements provide Web 2.0 sites with information-storage, creation, and *dissemination* challenges and capabilities that *go beyond* what the public formerly expected in the environment of the so-called "Web 1.0".

Web 2.0 Websites typically include some of the following features/techniques:
- Cascading Style Sheets to aid in the separation of presentation and content
- Folksonomies (collaborative tagging, social classification, social indexing, and social tagging)
- Microformats extending pages with additional semantics
- **REST** and/or XML- and/or **JSON**-based APIs
- Rich Internet application techniques, often Ajax and/or Flex/Flash-based
- Semantically valid XHTML and HTML markup
- Syndication, aggregation and notification of data in RSS or Atom feeds
- *Mashups*, merging content from different sources, client- and server-side
- Weblog-publishing tools
- Wiki or forum software, etc., to support user-generated content
- Internet privacy, the extended power of users to manage their own privacy in *cloaking* or deleting their own user content or profiles

Associated Innovations

It is a common misconception that "Web 2.0" refers to various visual design elements such as rounded corners or drop shadows. *While such design elements have commonly been found on popular Web 2.0 sites, the association is more one of fashion, a designer preference which became popular around the same time that "Web 2.0" became a buzz word.*

Another common misassociation with Web 2.0 is Ajax. This error probably *comes about* because many Web 2.0 sites rely heavily on Ajax or associated **DHTML** effects. So while Ajax is often required for Web 2.0 sites to function well, it is (usually) not required for them to function.

The Freemium business model is also characteristic of many Web 2.0 sites, with the idea that core basic services are given away for free, in order to build a large user base by *word-of-mouth* marketing. ***Premium*** service would then be offered for a price.

Criticism

The argument exists that "Web 2.0" does not represent a new version of the World Wide Web at all, but merely continues to use so-called "Web 1.0" technologies and concepts. Techniques such as Ajax do not replace underlying protocols like HTTP, but add an additional layer of abstraction on top of them. Many of the ideas of "Web 2.0" had already been featured in implementations on networked systems well before the term "Web 2.0" emerged. Amazon.com, for instance, has allowed users to write reviews and consumer guides since its launch in 1995, in a form of self-publishing. Amazon also opened its API to outside developers in 2002. Previous developments also came from research in computer-supported collaborative learning and computer-supported cooperative work and from established products like Lotus Notes and Lotus Domino.

In a podcast interview Tim Berners-Lee described the term "Web 2.0" as a "piece of jargon." "Nobody really knows what it means," he said, and went on to say that "if Web 2.0 for you is blogs and wikis, then that is people to people. But that was what the Web was supposed to be *all along.*"

Other criticism has included the term "a second ***bubble***" (referring to the Dot-com bubble of *circa* 1995–2001), suggesting that too many "Web 2.0" companies attempt to develop the same product with a lack of business models. The Economist has *written of* "Bubble 2.0."

Venture capitalist Josh Kopelman ***noted*** that Web 2.0 had excited only 530,651 people (the number of subscribers at that time to TechCrunch, a Weblog covering Web 2.0 matters), too few users to make them an economically ***viable*** target for consumer applications.

A few critics cite the language used to describe the ***hype*** cycle of Web 2.0 as an example of Techno-***utopianist rhetoric***. According to these critics, Web 2.0 is not the first example of communication creating a false, ***hyper-inflated*** sense of the value of technology and its impact on culture. The dot com ***boom*** and subsequent ***bust*** in 2000 was a ***culmination*** of rhetoric of the technological ***sublime*** in terms that would later make their way into Web 2.0 jargon. Communication as culture: essays on media and society (1989) and the technologies worth as represented in the stock market. Indeed, several years before the dot com stock market crash, the then-Federal Reserve chairman Alan Greenspan equated the *run up* of stock values as irrational ***exuberance***. Shortly before the crash of 2000 a book by Shiller, Robert J. Irrational Exuberance (Princeton, NJ: Princeton University Press, 2000.) was released detailing the overly optimistic ***euphoria*** of the dot com industry. The book Wikinomics: How Mass Collaboration Changes Everything (2006) even goes as far as to quote critics of the value of Web 2.0 in an attempt to acknowledge that hyper inflated expectations exist but that Web 2.0 is really different.

 Words

wiki[wiki] *n.*维基	**withhold**[wið'həuld] *v.*拒给，保留
folksonomy[fəuk'sɔnəmi] *n.*分众分类	**dissemination**[di,semi'neiʃən] *n.*分发
proliferation[prəu,lifə'reiʃən] *n.*激增，扩散	**mashup**['mæʃʌp] *n.* 糅合
embrace[im'breis] *v.*包括，利用	**cloak**[kləuk] *v.*掩饰，掩盖
abridge[ə'bridʒ] *v.*删节，精简	**premium**['primjəm] *adj.*质优的，高价的
encompass[in'kʌmpəs] *v.*构成，包括	**bubble**['bʌbl] *n.*泡沫，幻想的计划
leverage['li:vəridʒ] *v.*杠杆作用，好像通过杠杆作用进行影响	**circa**['sə:kə] adv.大约
	note[nəut] *v.*着重提到，表明，指出
agile['ædʒail] *adj.*敏捷的，灵活的	**viable**['vaiəbl] *adj.*切实可行的，可实施的
typify['tipifai] *v.*代表	**hype**[haip] *n.*大肆宣传，大做广告
syndication['sindikeiʃən] *n.*整合，聚合	**utopian**[ju:'təupjən] *adj.*乌托邦的，理想化的
embody[im'bɔdi] *v.*包含，体现	**rhetoric**['retərik] *n.*修辞，花言巧语
allude[ə'lju:d] *v.*暗指，间接提到	**hyper-**['haipə] *pref.*超出，过于
podcast['pɔdka:st] *n.*播客	**inflated**[in'fleitid] *adj.*夸张的
feed[fi:d] *n.*进给	**boom**[bu:m] *v.*兴隆，繁荣
column['kɔləm] *n.*专栏	**bust**[bʌst] *v.*破产，失败
silo['sailəu] *n.*筒仓，地窖	**culmination**[kʌlmi'neiʃn] *n.*顶点
perception[pə'sepʃən] *n.*感知，理解	**sublime**[sə'blaim] *n.*顶点
participatory[pɑ:'tisipeitəri] *adj.*供人分享的，提供参加机会的	**exuberance**[ig'zju:bərəns] *n.*生气勃勃
	euphoria[ju:'fɔ:riə] *n.*兴高采烈

 Phrases

ease of	解除，减少	**buzz word**	热门词语

pick-up	获得	come about	发生
in proportion as	按…比例，和…相称	word-of-mouth	口碑
give rise to	导致，使…发生	all along	连续，始终
free-ride	搭便车	write off	一口气写成
by way of	经由，作为	run up	上涨，兴起
go beyond	超出，超过		

 Abbreviations

RSS	Really Simple Syndication	聚合新闻服务
Ajax	Asynchronous JavaScript and XML	异步JavaSaript和XML
REST	Representational state transfer	表象化状态转换
JSON	JavaScript Object Notation	JavaScript标识对象的方法
DHTML	Dynamic HTML	动态HTML

Complex Sentences

1. Original: O'Reilly and John Battelle argued that the Web had become a platform, with software above the level of a single device, leveraging the power of the "Long Tail", and with data as a driving force.

Translation: O'Reilly和John Battelle指出，Web已经成为一个平台，该平台具有高于单一设备层的软件，大大影响了"长尾"力量，并且将数据做为驱动力。

2. Original: The impossibility of excluding group-members who don't contribute to the provision of goods from sharing profits gives rise to the possibility that rational members will prefer to withhold their contribution of effort and free-ride on the contribution of others.

Translation: 如果不能将那些不做贡献去提供有价值东西的小组成员排除在共享收益之外，就会导致这样的可能性：理性的成员宁愿保留他们的努力贡献并在其他人的贡献之上不劳而获。

Exercises

I. Read the following statements carefully, and decide whether they are true (T) or false (F) according to the text.

____1. O'Reilly classified Web 2.0 sites into four levels in the hierarchy according to the degree of the dependence of applications on Internet.

____2. Ajax is one of Internet application techniques which Web 2.0 sites often rely heavily on for featuring a rich, user-friendly interface.

____3. Web 2.0 adopts entirely new protocols and adds an additional layer of abstraction on top of them.

___4. Web 2.0 encourages users' contributions to Website content as well as supports their privacy.

___5. All services are given for free by the Web 2.0 sites using the Freemium business model.

II. Choose the best answer to each of the following questions.

1. What does the Eric Schmidt's abridged slogan, "*don't fight the Internet*" mean?
   (A) Don't reverse the interrelation of the Internet as a platform and the applications built on it.
   (B) Don't doubt the currently accepted technologies relating to the Internet.
   (C) Don't attack the Internet maliciously.
   (D) Don't attempt to create a new network competing with the Internet.

2. Which of the following is emphasized most about the criticism on Web 2.0 in this article?
   (A) Lacking the innovation of new technologies and concepts.
   (B) Being so ideal and abstract that users can't really knows what it means.
   (C) Tending toward an irrational exuberance.
   (D) Lacking users to become an economically viable target for consumer applications.

3. Which of the following is WRONG about the significance of Web 2.0?
   (A) Web 2.0 renovates and updates many technical specifications relating to the Web.
   (B) Web 2.0 changes the ways software developers and end-users utilize the Web.
   (C) Web 2.0 enhances creativity, information sharing, and collaboration among Web users.
   (D) Web 2.0 enables business to embrace the Web as a platform and use its strengths more effectively.

III. Translating.

1. According to Tim O'Reilly, Web 2.0 is the business revolution in the computer industry caused by the move to the Internet as platform, and an attempt to understand the rules for success on that new platform.

2. While such design elements have commonly been found on popular Web 2.0 sites, the association is more one of fashion, a designer preference which became popular around the same time that "Web 2.0" became a buzz word.

Part 2　Simulated Writing: Proposal

⊙Introduction

A proposal is a document written to persuade readers to follow a plan or course of action that you are proposing. Regardless of the type of proposal you will write, you must first consider the audience and purpose, the project management, and the proposal structure.

⊙Audience and Purpose

A proposal offers a plan to fill a need or requirement. It tells what you are proposing to do, how you plan to do it, when you plan to do it, and how much it will cost. Proposals often require more than one level of approval; so it must take various types of readers into account. You must consider their levels of technical knowledge of the subject and their position in the company.

The task of writing a persuasive proposal can be simplified by composing a concise statement of exactly the problem or opportunity that your proposal is designed to address. This helps you and your reader understand the value, scope and limitations of your proposal solution.

Unsolicited and Solicited Proposals

Unsolicited proposals are proposals submitted to a company without a prior request for a proposal. Companies often operate for years with a problem they have never recognized. You might prepare an unsolicited proposal if you were convinced that the potential customer could realize substantial benefits by adopting your solution to a problem. You should need to convince the customer of the need for what you are proposing and that your solution would be the best one.

Solicited proposals are proposals prepared in response to a request for goods or services. Procuring organizations that would want companies to compete for a job commonly issue a Request for Proposals (RFP). Often an RFP will define a problem and allow those who respond to suggest possible solutions. Your first task in writing a proposal is to find out exactly what your prospective customer needs. To do that, survey your potential customer's business, and determine whether your organization can satisfy the customer's needs. Before preparing a sales proposal, try to find out who your principal competitors are. Compare your company's strengths and advantages with those of the competing firms. Determine your advantages over your competitors, and emphasize those advantages in your proposal.

Proposal Structure and Contents

Depending on the scope and formality of the proposal, a typical outline may include the following sections:

1. Front Matter

Cover Letter or Letter of Transmittal. In the cover letter, express your appreciation for the opportunity to submit your proposal, any assistance the customer has provided, and any previous positive associations. Summarize the proposal's recommendations and express confidence that your company will satisfy the customer's needs.

Title Page. Include the title of the proposal, the date, the organization to which it is being submitted , and your company name.

Table of Contents. Include a table of contents in longer proposal to guide readers to important headings, which should be listed according to beginning page numbers.

List of Figures. If your proposal has six or more figures, include a list of figures to help your reader to relevant material.

2. Body

Executive Summary. Briefly summarize the proposal's highlights in non-technical, easy to understand language.

Introduction. Explain the reasons for the proposal. Discuss your understanding of the problem. Discuss any assumptions made. Also state the purpose and scoped of your proposal.

Background or Problem Statement. This section should describe the problem you propose to solve and your solution to it. It should also indicate the dates on which you propose to deign and complete work. Most importantly, it should state any special benefits of your proposed approach, and the cost of the project.

Product Description. If your proposal offers products as well as services, include a general description of the products and any technical specifications.

Detailed Solutions. Explain in a detailed section that will be read by technical specialists exactly how you plan to do what you are proposing and explain the product or service you are offering.

Cost Analysis. Itemize the estimated costs of all the products and services you are offering.

Delivery Schedule. Outline how you will accomplish the work and show a timetable for each phase of the project.

Staffing. Summarize the expertise (education, experience, and certifications) of key personnel and include their resumes as an appendix.

Training Requirements. If the products and services you are proposing require training the customer's employees, specify the required training and its cost.

Organization. Describe your company, its history, and its present position in the industry. Promote the company. Conclude the proposal on an upbeat persuasive note.

Conclusion. Include a persuasive conclusion that summarizes the proposal's key points and stresses your company's strong points. The conclusion should persuasively resell your proposal by emphasizing the benefits of your solution, product, or service over any competing ideas. Also include details about the time period during which the proposal is valid. Show confidence in your solution, your appreciation for the opportunity to submit the proposal, and your willingness to provide further information.

3. Back Matter

Appendixes. Provide material of interest to specialized readers — technical graphics, statistical analysis, charts, tables, or sources of further information.

Glossary. If your proposal contains terms that will be unfamiliar to your intended audience, list their definition in a glossary.

Sample

Software and Hardware Proposal
for Technical Communications Department

I. Executive Summary

The purpose of this proposal is to define the computer hardware and software requirements for outfitting the startup Technical Communications Department.

The Technical Communications Department will be responsible for preparing consumer end user and administrator documentation in support of our software product. Documentation will be provided to the customer in printed form, as well as on CD, and additional documentation support will be available via Web-based material.

In order to determine the proper mix of hardware and software to support this effort, an online survey was conducted among software technical writers. Suggestions for hardware requirements and software preferences were collected, and the results evaluated against the focus of the Technical Communications Department. A draft hardware and software configuration was developed, and prices were researched. The suggested cost breakdown is included in Section V.

II. Technical Description

Overview

The Technical Communications Department will expand to a staff of ten (10) technical writers. Additional specialties in Web design and graphics design will be incorporated within the group to support the overall effort.

Ten (10) standard workstations will be provided for the day-to-day technical writing duties, with a separate, more powerful workstation for graphics software and specialized peripherals. This workstation can also act as a backup unit in case of failure on one of the primary units.

Peripherals specified for this department include a HP Deskjet color inkjet printer for draft color materials. In addition, a production Lexmark laser printer will be used for day-to-day printing. A color scanner will also be provided for use in preparing graphics for the various editions of documentation. A CD-RW unit will be provided for creating original CD-ROM materials for reproduction.

Software for the department will include Microsoft Office 97 Professional Edition for all workstations. This will provide basic word processing, spreadsheet, presentation, and data base functionality. Adobe Framemaker will be used to prepare formatted documentation text for publication. Macromedia Dreamweaver with Allaire's HomeSite bundled will be used to prepare Web pages.

Hardware Details

A. Standard Workstations (10)

The standard workstation for this department will be a Dell Dimension XPS T, averaging under $3300 per unit, including software. Each standard unit will be a Pentium III 450 Mhz PC with MS Office 97 Professional loaded and

- 9.1Gb hard drive
- 96Mb RAM
- 40X CD-ROM
- 8Mb video card
- 19" monitor
- Microsoft Windows '98
- Microsoft Office 97 Professional Edition
- Adobe Framemaker
- Macromedia Dreamweaver with Allaire HomeSite 4.0 bundled

B. Graphics Workstation/Backup Unit (1) and Peripherals

The graphics workstation with peripherals will include all of the above, but will include 21" monitor, 128Mb RAM and a 16Mb RAM video card in support of graphics applications and peripherals:

- HP ScanJet Flatbed Scanner
- CD-RW

The cost of the base system will be around $4300. By bundling the scanner and CD-RW, along with the HP Deskjet printer (noted below), we may achieve some savings.

C. Printers

Printers selected for this department include:
- HP Deskjet 895Cse Color Printer and cable
- Lexmark Optra S 1855 black and white production printer and cable

Total Budget

The overall budget for this department is $38,708.45 (see Section V for breakdowns).

III. Technical Requirements

Assistance from the Purchasing Department will be needed to locate the best prices and delivery schedules for software.

Support from the IT Department will be needed to set up each workstation and integrate it into the company LAN. The IT Department will also take responsibility for processing licenses, warranties, and software updates.

IV. Business Issues

The computer equipment will be purchased with a three-year warranty and three year premium next business day service contract.

Formal training on the software has not been included in this proposal. It is assumed that each technical writer will have some familiarity with the software associated with his/her job function. If necessary, internal training may be done by staff members who show a particular aptitude in the application. If no one can conduct this training and it proves to be necessary, it may be necessary to arrange for outside training.

V. Total Cost Matrix

Qty	Description	Per Unit	Extended
10	Standard Workstations Office Professional	$1,995.95	$19,959.50
1	Graphics/Backup Workstation Office Professional, scanner, CD-RW	$3,021.95	$3,021.95
12	Framemaker	$895.00	$10,740.00
12	Dreamweaver 2 HomeSite 4.0	$299.00	$3,588.00
1	Lexmark Printer	$1,000.00	$1,000.00
1	HP Deskjet 895Cse Color Printer	$399.00	$399.00
	Total		$38,708.45

[The Front Matter and the Back Matter are omitted from this proposal.]

Part 3   Listening and Speaking

⧉ Dialogue: Enhancing Your Computer Security

(For effectively fending off the malicious intruders away from the computer, Henry, Mark and Sophie are having a discussion to share their knowledge and experience about the basic strategies for enhancing the computer security.)

Henry	In my opinion, one of the most ordinary methods is to use *anti-virus* software and firewalls.[1]
Mark	So, to keep the computer safe, we can set it at the highest security level when configuring all anti-virus software and firewalls, right?
Henry	Only half right, Mark. Higher security level is recommended, but it shouldn't block your normal Web browsing. To handle this problem, you can **layer** your security and make sure to enable the security settings on Web browsers and disable file sharing.
Sophie	Further more, I think it's very important to **patch** and upgrade regularly. We should get the latest updates and patches for the operating system, security software and Web browsers as soon as they become available.
Henry	That's right! Similarly, we should **cultivate** the habit of backing up and encrypting the important and sensitive data on our computer.[2]
Mark	Besides the threats to security from the computer itself, there may other threats within the network.
Sophie	So, how to protect multiple computers linked together at home or at the office?
Mark	The administrator should *be charged with* the task of checking to make sure there are no strangers **loitering** or causing trouble to increase network security. If there are, ensure that any infected computers are removed from the network and disinfected as soon as possible.[3]
Henry	And then, what should be **cautioned** when we connect to the Web?
Sophie	Strengthen the passwords primarily. It is recommended that passwords should have at least eight characters and should combine alphanumeric and special characters, such as $, *, and &. Moreover, we are supposed to change our passwords every 45-60 days to enhance security.[4]

Mark	Besides, another threat from the Web may come from emails. Sometimes, you may think an email is harmless until you open it and it's not. So **beware** of suspicious email.
Sophie	But sometimes we may receive a mass of emails at one time, how do we distinguish them from the collection?
Mark	Scan emails before you open them. Take extra caution when **launching** executable (.exe) files attached to an email and never open attachments in emails from unknown senders. Never reply to **spam** and use anti-spam software whenever possible.[5]
Henry	Never click and open **unauthentic** Websites as well.[6]
Sophie	Ah, so many ideas we've reviewed, and all of them should be considered **comprehensively** in practice as basic strategies for enhancing the computer security.
Henry & Mark	That's right!

Exercises

Work in pairs, and make up a similar conversation by replacing the scenarios with other material in the below.

Reducing Spam and Boosting Your Email Security

[1] Configure your anti-virus software to automatically scan your incoming emails for viruses. email is the favorite medium of distributing malicious software. Make sure you keep your anti-virus software definitions *up to date*.

[2] If you are interested in signing up for free stuff on the Internet, always start using a separate e-mail account just for this purpose. Many free account providers like Hotmail, Yahoo!, Gmail are available.

[3] While posting email addresses in a blog use a format that is only recognizable to a human, like ABC at hotmail.com instead of ABC@hotmail.com. This will reduce automatic **bots** from harvesting your email from the blog to the **spammers**.

[4] Do not open an email from an unverified source. Delete it, especially if it comes with an attachment. Never ever reply to these emails, since a reply would only tell the spammer that your email is "alive".

[5] If you get an official message from your bank or eBay or another site you are not sure it is *genuine* here is what you do. Instead of clicking on the link embedded in the mail log on to the site normally via your browser. If there are any genuinely serious problems you should get a message when you log on. Alternatively contact the site's customer service via the phone if possible.

[6] Consider using standalone spam filtering software. This software analyses your emails for common characteristics of spam emails including words like "click". It also compares senders' emails against a "Friends List."

Words

layer['leiə] *v.*把…形成或分成层次	**spam**[spæm] *n.*垃圾邮件
patch[pætʃ] *v.*修补，打补丁	**unauthentic**['ʌnɔ:'θentik] *adj.*不可靠的，不可信的
cultivate['kʌltiveit] *v.*培养，养成	**comprehensively**[ˌkɔmpri'hensivli] *adv.*全面地
loiter['lɔitə] *v.*闲逛，徘徊	**bot**[bɔt] *n.*机器人程序
caution['kɔ:ʃən] *v.*警告	**spammer**[spæmə] *n.*发垃圾邮件的人
beware[bi'wɛə] *v.*小心，谨防	**genuine**['dʒenjuin] *adj.*真实的，真的
launch[lɔ:ntʃ] *v.*发动，开始	

Phrases

fend off	挡开，躲开	**be charged with**	负…责任
anti-virus	防（计算机）病毒	**up to date**	最新的，直到最近的

Listening Comprehension: History of Google

Listen to the passage and the following 3 questions based on it. After you hear a question, there will be a break of 10 seconds. During the break, you will decide which one is the best answer among the four choices marked (A), (B), (C) and (D).

Questions

1. When was Google Inc. set up?
2. What is the fundamental idea of Google search engine in technology?
3. Why does the author cite the fact that the verb "to google" was officially added to two authoritative dictionaries?

Choices

1. (A) 1996   (B) 1997   (C) 1998   (D) 1999

2. (A) Ranking results according to the profitability of advertising

(B) Ranking results based on the analysis of the relationships between Web sites

(C) Ranking results according to the number of times the search term appeared on a page

(D) Ranking results based on the position the search term appeared on a page

3. (A) Demonstrating that Google has become quite well known and has most powerful influence on modern life.

(B) Demonstrating that the results searched out by Google are increasingly supplementing those two dictionaries.

(C) Demonstrating that Google provides online advertising service for those two dictionaries.

(D) Demonstrating that Google search engine can search out every word in those two dictionaries.

 Words

hypothesize[hai'pɔθisaiz] *v.*假设，猜测	**following**['fɔləuiŋ] *n.*崇拜者，追随者
backlink[bækliŋk] *n.*反向链接	**outsize**['aut,saiz] *adj.*特大的，极广阔的
domain[dəu'mein] *n.*域名	

⫸ Dictation: How Web Search Engines Work

This passage will be played THREE times. Listen carefully, and fill in the blanks with the words you have heard.

A Web search engine is a search engine designed to search for information on the World Wide Web. Information may _____1_____ of Web pages, images and other types of files. Unlike Web directories, which are maintained by human editors, search engines _____2_____ *algorithmically* or are a _____3_____ of algorithmic and human input.

Web search engines work by storing information about many Web pages in the following _____4_____: Web crawling, indexing and searching. They _____5_____ these pages from the WWW itself by sending out a *Web crawler* (sometimes also known as a _____6_____) — an _____7_____ Web browser which follows every link it sees. Data about Web pages are collected, *parsed,* and stored by another program, called an indexer which creates an index based on the contents of each page. The purpose of storing an index is to _____8_____ speed and performance in finding relevant pages for a search _____9_____. Without an index, the search engine would scan every page in the *corpus*, which would require _____10_____ time and computing power.

Some search engines, such as Google, store all or part of the _____11_____ page, referred to as a _____12_____, as well as information about the Web pages, whereas others, such as

AltaVista, store every word of every page they find. This cached page always _____13_____ the actual search text since it is the one that was actually indexed, so it can be very useful when the content of the current page has been _____14_____ and the search _____15_____ are no longer in it. This problem might be considered to be a mild form of *LinkRot*, and Google's handling of it increases usability by _____16_____ user expectations that the search terms will be on the _____17_____ Web page. This satisfies the principle of least *astonishment* _____18_____ the user normally expects the search terms to be on the returned pages. Increased search _____19_____ makes these cached pages very useful, even _____20_____ the fact that they may contain data that may no longer be available elsewhere.

Words

algorithm['ælgəriðəm] *n.*算法	**corpus**['kɔːpəs] *n.*文集，（事物的）主体
parse[pɑːz] *v.*分解，解析	**astonishment**[əs'tɔniʃmənt] *n.*惊讶

Phrases

Web crawler	网络爬虫，爬网程序	**LinkRot**	出错链接页面

E-Commerce

Unit

9

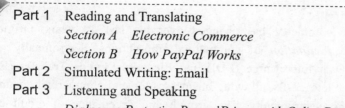

Part 1   **Reading and Translating**

≫ **Section A   Electronic Commerce**

Electronic commerce, commonly known as e-commerce or eCommerce, consists of the buying and selling of products or services over electronic systems such as the Internet and other computer networks. The amount of trade conducted electronically has grown extraordinarily since the spread of the Internet. *A wide variety of commerce is conducted in this way, spurring and drawing on innovations in electronic funds transfer, supply chain management, Internet marketing, online transaction processing, electronic data interchange (**EDI**), inventory management systems, and automated data collection systems. Modern electronic commerce typically uses the World Wide Web at least at some point in the transaction's lifecycle, although it can encompass a wider range of technologies such as e-mail as well.*

A large percentage of electronic commerce is conducted entirely electronically for virtual items such as access to premium content on a Website, but most electronic commerce involves the transportation of physical items in some way. Online retailers are sometimes known as e-tailers and online retail is sometimes known as e-tail. Almost all big retailers have electronic commerce presence on the World Wide Web.

Electronic commerce is generally considered to be the sales aspect of e-business. It also consists of the exchange of data to facilitate the financing and payment aspects of the business transactions.

Early Development

The meaning of electronic commerce has changed over the last 30 years. Originally, electronic commerce meant the *facilitation* of commercial transactions electronically, using technology such as Electronic Data Interchange (EDI) and Electronic Funds Transfer (**EFT**). These were both introduced in the late 1970s, allowing businesses to send commercial documents like purchase *orders* or *invoices* electronically. The growth and acceptance of credit cards, automated teller machines (**ATM**) and telephone banking in the 1980s were also forms of electronic commerce. From the 1990s onwards, electronic commerce would additionally include enterprise resource planning systems (**ERP**), *data mining* and *data warehousing*.

Perhaps it is introduced from the Telephone Exchange Office, or maybe not. The earliest example of many-to-many electronic commerce in physical goods was the Boston Computer Exchange, a marketplace for used computers launched in 1982. The first online information marketplace, including online consulting, was likely the American Information Exchange, another pre-Internet online system introduced in 1991.

Although the Internet became popular worldwide in 1994, it took about five years to introduce security protocols and DSL allowing continual connection to the Internet. And by the end of 2000, a lot of European and American business companies offered their services through the World Wide Web. *Since then people began to associate a word "ecommerce" with the ability of purchasing various goods through the Internet using secure protocols and electronic payment services.*

Business Applications

Some common applications related to electronic commerce are the following:
- E-mail and *messaging*
- Content management systems
- Documents, spreadsheets, database
- Accounting and finance systems
- Orders and *shipment* information

- Enterprise and client information reporting
- Domestic and international payment systems
- Newsgroup
- On-line shopping
- Messaging
- Conferencing

Forms

Contemporary electronic commerce involves everything from ordering "digital" content for immediate online consumption, to ordering conventional goods and services, to "meta" services to facilitate other types of electronic commerce.

On the consumer level, electronic commerce is mostly conducted on the World Wide Web. An individual can go online to purchase anything from books, grocery to expensive items like real estate. Another example will be online banking like online bill payments, buying stocks, transferring funds from one account to another, and initiating wire payment to another country. All these activities can be done with a few *keystrokes* on the keyboard.

On the institutional level, big corporations and financial institutions use the Internet to exchange financial data to facilitate domestic and international business. Data integrity and security are very hot and *pressing* issues for electronic commerce these days.

Business-to-Business

Electronic commerce that is conducted between businesses is referred to as Business-to-Business or **B2B**, as opposed to those between businesses and other groups, such as business and individual consumers (**B2C**) or business and government (**B2G**). B2B can be open to all interested parties (eg. commodity exchange) or limited to specific, pre-qualified participants (private electronic market).

B2B can also refer to all marketing activities between businesses made in an industry value chain not just the final transactions that result from marketing.

The volume of B2B transactions is much higher than the volume of B2C transactions. One reason for this is that businesses have adopted electronic commerce technologies in greater numbers than consumers. Also, in a typical supply chain there will be many B2B transactions but only one B2C transaction, as the completed product is retailed to the end customer. An example of a B2B transaction is a chicken feed company selling its product to a chicken farm, which is another company. An example of a B2C transaction is a grocery store selling grain-fed chickens to a consumer.

⊖ E-Marketplace

"E-" or "electronic" marketplace in a B2B context is primarily a large online platform (B2B portal) or Website that facilitates interaction and/or transactions between buyers and suppliers at organizational or institutional rather than individual levels. Since the builders of such marketplaces primarily aim at facilitating buyer-seller interaction (in most cases without being a buyer or seller themselves), these are also referred to as "third-party" B2B marketplaces.

These marketplaces can do one or more of the following:

- Help buyers find new suppliers and *vice versa*
- Help reduce the time and cost of interaction for B2B transactions
- Help increase trade between distant geographies and cultures
- Help manage payments and track orders for B2B transactions
- Help the environment by using appropriate technology that is environmentally friendly

1. Vertical e-Marketplace

A vertical e-marketplace or Vortal *spans up and down* every segment of one specific industry. Each level of the industry has access to every other level, which greatly increases collaboration. Buyers and sellers in the industry are connected to increase operating efficiency and decrease supply chain costs, inventories and cycle times. This is possible because buying/selling items in a single industry standardizes needs, thereby reducing the need for *outsourcing* many products.

2. Horizontal e-Marketplace

A horizontal e-marketplace connects buyers and sellers across many industries. The most common type of materials traded horizontally is **MRO** (maintenance, repair and operations) materials. Mainly business and consumer *articles*, these items are in demand because they are crucial to the daily running of a business, regardless of industry and level within that industry. Many corporations have MRO materials bought directly on-line by the maintenance team in order to *relieve* the purchasing department.

3. No-frills e-Marketplace

Developed in response to customers wanting to purchase products without service (or with very limited service), the no-*frills* e-marketplace *parallels* the B2C offering of no-frills budget airlines. The subject of several Harvard and **IMD** articles/case-studies, no-frills B2B e-marketplaces enable the effective de-bundling of service from product via clear "business rules." This provides the basis of differentiation from conventional B2B sales/purchasing channels.

Words

facilitation[fə,sili'teiʃən] *n.*简易化	**span**[spæn] *v.*跨越
order['ɔ:də] *n.*订单	**outsource**['aut,sɔ:s] *v.*外界供应，外包
invoice['invɔis] *n.*发票，发货单	**article**['ɑ:tikl] *n.*商品，项目
message['mesidʒ] *v.*即时通讯	**relieve**[ri'li:v] *v.*为…提供帮助或援助
shipment['ʃipmənt] *n.*运送，运输	**frill**[fril] *n.*装饰
keystroke['ki:strəuk] *n.*按键	**parallel**['pærəlel] *v.*匹配，与…相应
pressing['presiŋ] *adj.*紧迫的，迫切的	

 Phrases

draw on	引起，利用	**vice versa**	反之亦然
data mining	数据挖掘	**up and down**	到处，详细
data warehouse	数据仓库		

 Abbreviations

EDI	Electronic Data Interchange	电子数据交换
EFT	Electronic Funds Transfer	电子资金转账
ATM	Automated Teller Machines	自动柜员机
ERP	Enterprise Resource Planning	企业资源规划
B2B	Business-to-Business	企业对企业的电子商务模式
B2C	Business-to-Customer	企业对消费者的电子商务模式
B2G	Business-to-Government	企业对政府的电子商务模式
MRO	Maintenance, Repair and Operations	维护、维修、运行
IMD	International Institute for Management Development	国际管理发展研究院

 Complex Sentences

1. **Original:** Modern electronic commerce typically uses the World Wide Web at least at some point in the transaction's lifecycle, although it can encompass a wider range of technologies such as e-mail as well.

Translation: 通常来说现代电子商务至少在交易生命周期的某些方面利用万维网，尽管它也能包含诸如电子邮件这样的更广的技术。

Exercises

I. Read the following statements carefully, and decide whether they are true (T) or false (F) according to the text.

___1. Modern electronic commerce exclusively uses the World Wide Web through the transaction's lifecycle.

___2. Most electronic commerce involves the transportation of physical items, but it can be conducted entirely electronically for virtual items as well.

___3. Boston Computer Exchange is the first online information marketplace.

___4. The greater numbers of e-commerce technologies adopted in businesses result in the much higher volume of B2B transactions than B2C transactions.

___5. For the institutional level of electronic commerce, data integrity and security are very hot and pressing issues these days.

II. Choose the best answer to each of the following questions.

1. Which of the following technologies related to ecommerce is introduced last?

(A) Automated teller machines

(B) Electronic data interchange

(C) Enterprise resource planning

(D) Electronic Funds Transfer

2. Which of the following is RIGHT about the comparison between B2B and B2C?

(A) Multiple B2B transactions are involved in a typical supply chain but only one B2C transaction is.

(B) B2B is limited to one specific industry in one supply chain while B2C can cross many industries.

(C) B2B just involves transactions made in an industry value chain while B2C refers to the final transactions that result from marketing.

(D) B2B is limited to specific, pre-qualified business participants while B2C is open to all interested individual consumers.

3. Which of the following is WRONG about the E-marketplace?

(A) E-marketplace is primarily a large online platform or Website that facilitates interaction and/or transactions between buyers and suppliers.

(B) A vertical e-Marketplace connects buyers and sellers in the industry to increase operating efficiency and decrease supply chain costs, inventories and cycle times.

(C) A horizontal e-marketplace connects buyers and sellers across many industries for items which are crucial to the daily running of a business.

(D) A no-frills e-marketplace enables the effective de-bundling of service from product, bypassing business rules.

III. Translating.

1. A wide variety of commerce is conducted in this way, spurring and drawing on innovations in electronic funds transfer, supply chain management, Internet marketing, online transaction

processing, electronic data interchange (EDI), inventory management systems, and automated data collection systems.

2. Since then people began to associate a word "ecommerce" with the ability of purchasing various goods through the Internet using secure protocols and electronic payment services.

▶ Section B　How PayPal Works

The simple idea behind PayPal – using encryption software to allow people to make financial transfers between computers – has turned into one of the world's primary methods of online payment. Despite its occasionally troubled history, including fraud, lawsuits and *zealous* government *regulators*, PayPal now boasts more than 100 million accounts worldwide.

PayPal is an online payment service that allows individuals and businesses to transfer funds electronically. You can use it to pay for online auctions, purchase goods and services, or to make donations. You can even use it to send cash to someone.

A basic PayPal account is free. You can send funds to anyone with an e-mail address, whether or not they have a PayPal account. They'll get a message from PayPal about the funds, and then they just have to *sign up* for their own account.

Funds transferred via PayPal reside in a PayPal account until the holder of the funds retrieves them or spends them. If the user has entered and verified their bank account information, then the funds can be transferred directly into their account.

◉ PayPal Infrastructure

PayPal doesn't fundamentally change the way merchants interact with banks and credit card companies. It just acts as a middleman. Credit and *debit card* transactions travel on different networks. When a merchant accepts a charge from a card, the merchant pays an interchange, which is a small fee of about ten cents plus approximately 2 percent. The interchange is made up of a variety of small fees paid to all the different companies that have a part in the transaction -- the merchant's bank, the credit card association and the company that issued the card. If someone pays by *check*, a different network is used, one that costs the merchant less but moves more slowly.

What part does PayPal play in all this? Both buyer and seller *deal with* PayPal, having already provided their bank account or *credit card* information. PayPal, in turn, handles all the transactions with various banks and credit card companies, and pays the interchange. They make this back on the fees they charge for receiving money, as well as the interest they collect on money left in PayPal accounts.

PayPal touts its presence as an extra layer as a security feature, because everyone's information, including credit card numbers, bank account numbers and address, stays with PayPal. With other online transactions, that information is transmitted from the buyer to the merchant to the credit card processor.

PayPal also offers a $5 PayPal Security Key – a portable device that creates a six-digit code every 30 seconds. The user links this key to his or her eBay or PayPal account. The six-digit code is used in conjunction with the user ID and password to create a unique security code.

All the money held in PayPal accounts is placed into one or more bank accounts, where PayPal collects interest. Account holders do not receive any of the interest gained on their money. *Some PayPal critics claim that one of the reasons PayPal locks accounts and puts people through a long, **frustrating** appeal process is that they can keep the funds in the bank longer to collect more interest.*

⊜PayPal History

Peter Thiel and Max Levchin founded PayPal in 1999 under the name Confinity. The idealistic vision of the company was one of a borderless currency free from governmental controls. However, PayPal's success quickly drew the attention of **hackers**, *scam artists* and organized crime groups, who used the service for frauds and money **laundering**. New security measures *stemmed* the *tide* of fraud and customer complaints, but government officials soon *stepped in*. Regulators and attorney *generals* in several states, including New York and California, fined PayPal for violations and investigated the company's business *practices*. Some states, such as Louisiana, banned PayPal from operating in their states altogether. PayPal has since received licenses that allow them to operate in these places.

Despite the initial turmoil, PayPal's market share continued to grow. At first PayPal offered new users $10 to join, plus bonuses for referring friends. The service grew so quickly that it soon became the default online payment service. Buyers wanted to use it since so many merchants accepted it, and merchants accepted it because so many buyers were using it. PayPal *owes* much of its initial growth *to* eBay users who used the service to pay for items and accept payments for their online auctions. PayPal even beat eBay at the online payment business, *trumping* eBay's in-house payment system Billpoint so thoroughly that eBay bought PayPal in 2002. Then it *phased out* Billpoint and integrated PayPal into its services. Sellers with PayPal accounts can place icons in their auctions and buyers can simply click on a PayPal logo when they win an auction to make an immediate payment.

In early 2002, PayPal held its **IPO**, opening at $15.41 per share and closing the day's trading above the $20 *mark*. eBay purchased PayPal that same year for $1.4 billion in stock. Recently, eBay spent another $370 million to *buy out* another PayPal competitor, VeriSign.

Using PayPal: Sending Funds

More than 70 percent of all eBay sellers offer PayPal as a payment option, and a large chunk of PayPal's business still comes from online auctions. However, one of the keys to PayPal's success has been its ability to expand beyond the eBay market. You can use it to send money to a friend, donate to a charity and buy items from online merchants.

If you want to donate to a charity using PayPal, the process is just like sending money to anyone else. You need the charity's email address, or they might have a *button* on their Website that allows you to make a donation directly. The main difference lies in the "Category of Purchase" entry on the PayPal payment page. Technically, this would be a *quasi*-cash transaction. However, such a transaction could *be subject to* fees, depending on the source of the money – if you draw your PayPal funds from a credit card, you might be charged cash advance fees. You can just as easily select "Service" as the category, and the donation will work with no problems or fees.

You can use PayPal to purchase goods from non-eBay merchants who have set up a PayPal *storefront*. Once you've selected your items, go to the Website's *checkout* page. You will have the option of selecting a credit card or PayPal to pay for your purchase. Selecting PayPal may send you to a login page for your PayPal account. There you can transfer the appropriate amount to the merchant, who will then complete the sale. Some merchants integrate PayPal into the Website, meaning that you put your PayPal information directly into their site.

If a Website only accepts credit cards, you can still use funds in your PayPal account to make a purchase. PayPal users can use the "PayPal Debit Bar" to get a virtual *MasterCard* number. You can use that card number with any merchant who accepts MasterCard, and the funds will be deducted from the PayPal account. This service is free.

For example, you might want to use your PayPal account to buy something from Amazon.com. However, Amazon doesn't accept PayPal as a payment *method*. You can activate the Debit Bar from within your PayPal account. Assuming you are carrying enough of a *balance* in your account to cover the purchase, PayPal will give you a 16-digit number, just like a credit card number. Then you will select MasterCard as your payment method from Amazon's payment page and enter the Debit Bar number.

Using PayPal: Receiving Funds

Merchants who want to use PayPal to accept payments have a wide range of options available.

For basic payments, such as online auctions or simple Web site sales, the merchant can simply provide buyers with their e-mail address, and buyers can make the appropriate payments to the merchant's PayPal account. eBay sellers can place PayPal buttons on their auctions, and the checkout invoice PayPal sends to auction winners will include a link to pay via PayPal.

PayPal also provides extensive services for online merchants. Prior to services like PayPal, someone who wanted to accept credit card payments online had to set up a merchant account through a credit card company. Creating a Web interface to use this account could be confusing and difficult. PayPal bypasses this problem. Business or Premier PayPal accounts can set up a Buy Now button, a PayPal shopping cart, or options for ongoing subscriptions and *recurring* payments.

A Buy Now button allows merchants to paste a small piece of HTML code into their site, creating a button for buyers to click when they want to purchase an item. This takes the buyer to a secure payment page, where they enter their credit card information and shipping address. Once the transaction is complete, the money, minus PayPal's fees, is transferred directly into the merchant's account.

The PayPal shopping cart is more involved, but it has the same result. HTML code for various buttons (add to cart, view cart) is added to lists of items, and the item details are added by the merchant. Buyers can add the items they want to purchase to their cart, and when they *check out*, they'll go to a secure payment page, just like a Buy It Now page.

PayPal's two main merchant account types, Standard and Pro, offer slightly different packages. With a standard account, when a customer checks out at the shopping cart page, they go to the PayPal site to log in and make the payment. With a Pro account, PayPal processes the transaction in the background – the customer makes the entire sale on the merchant's site. A Pro account has higher percentages on transactions (2.2 to 2.9 percent versus 1.9 to 2.9 percent) and a $20 monthly fee. It also requires knowledge of Web services and APIs (Application Program Interfaces), as well as a minimum of two days for installation.

PayPal also *streamlines* transactions for merchants who sell to international users. It can convert funds to whatever currency the merchant wants for 2.5 percent.

 Words

zealous ['zeləs] *adj.* 热心的，积极的	**practice** ['præktis] *n.* 业务，常规工作
regulator ['regjuleitə] *n.* 管理者	**trump** [trʌmp] *v.* 胜过
check [tʃek] *n.* 支票	**mark** [mɑːk] *n.* 标志值
frustrate [frʌs'treit] *v.* 挫败，使感到灰心	**button** ['bʌtn] *n.* 按钮

hacker['hækə] n.电脑黑客	quasi-['kweisai] adj.类似，准，半
scam[skæm] n.& v.骗局，欺诈	storefront[stɔ:frʌnt] n.店头，店面
artist['ɑ:tist] n.骗子，家伙	checkout['tʃekaut] n.结算
launder['lɔ:ndə] v.洗黑钱	method['meθəd] n.办法，方法
stem[stem] v.抵抗，阻止	balance['bæləns] n.收支差额，结余
tide[taid] n.高潮	recur[ri'kə:] v.再发生，重现
general['dʒenərəl] n.（做事的）头儿，高级官员	streamline['stri:mlain] v.使简单化

Phrases

sign up	经报名（或签约）获得	phase out	使逐步淘汰，逐渐停止
debit card	借记卡	buy out	买下全部
deal with	与…打交道（做买卖）	be subject to	附属于，易受…的
credit card	信用卡	MasterCard	万事达信用卡
step in	介入，走进	check out	结账
owe to	把…归功于		

Abbreviations

IPO	Initial Public Offering	首次公开发行股票

Complex Sentences

1. **Original:** Some PayPal critics claim that one of the reasons PayPal locks accounts and puts people through a long, frustrating appeal process is that they can keep the funds in the bank longer to collect more interest.

   Translation: 一些PayPal的批评者宣称，PayPal锁定账户并让人们经受一个长期的、沮丧的申诉过程的原因之一是他们就能够将资金保存在银行中更长时间以获得更多的利息。

Exercises

I. Read the following statements carefully, and decide whether they are true (T) or false (F) according to the text.

____1. PayPal has fundamentally changed the way merchants interact with banks and credit card companies.

____2. Frauds and money laundering troubled PayPal in its initial growth.

____3. PayPal provides two main merchant account types, Standard and Pro, to adapt to different users.

____4. Buyers can put multiple items they want to purchase to the shopping cart before checking out.

___5. PayPal was integrated into eBay's services in 2002 and is only available within the eBay market.

II. Choose the best answer to each of the following questions.

1. Which of the following factors primarily made PayPal succeed in its initial growth?
   (A) PayPal attracts users by offering new users bonuses for joining or referring friends.
   (B) The development of online auctions requires a service used for online payments.
   (C) PayPal provides most of its services free.
   (D) PayPal stems any fraud and customer complaints effectively.
2. Which of the following is RIGHT about the services provided by PayPal?
   (A) PayPal supports funds transfer between the users neither of whom has PayPal accounts.
   (B) PayPal Security Key randomly creates a unique seaurity code for a user's eBay or PayPal account.
   (C) PayPal only allows for transfering funds electronically, not in the form of cash.
   (D) PayPal handles transactions with various banks and credit card companies, and pays the interchange.
3. Why does the author refer to the example of "*Amazon*"?
   (A) To prove that PayPal can be used to purchase goods from non-eBay merchants who have set up a PayPal storefront.
   (B) To indicate that Amazon is one of the biggest worldwide competitors of eBay.
   (C) To argue that not all the online shops accept PayPal as their payment method.
   (D) To show that PayPal enables users to still use funds in their PayPal account to make a purchase even if a Website only accepts credit cards.

III. Translating.

1. The simple idea behind PayPal – using encryption software to allow people to make financial transfers between computers – has turned into one of the world's primary methods of online payment.
2. A Buy Now button allows merchants to paste a small piece of HTML code into their site, creating a button for buyers to click when they want to purchase an item.

Part 2 Simulated Writing: Email

⊙ Introduction

Email is widely used in many settings. Although it is often seen as less formal than printed business letters, it may be faster and more efficient. Email allows individuals and groups to communicate with one another rather than face-to-face meetings or repetitive telephone calls, such as information exchange, brainstorming and problem solving, record keeping, group work, staying in touch professionally or socially, and transmitting documents. Effective email requires a clear sense of the purpose for writing, as well as a clear statement of the message.

⊙Common Sections

1. To, Cc & Bcc

Always make sure you e-mail the right people, in the right way.

The **To:** field is for people that the message directly affects, and that you require action from.

The **Cc:** field (or carbon copy) is mainly for people that do not need to act or reply to the message, but to keep informed of the content.

Finally, the **Bcc:** field (Blind Carbon Copy) is for people to receive the message without any of the other recipients knowing. Also used for an e-mail to hundreds of people.

2. Subjects

The Subject Line is one of the most important parts of an e-mail. The subject line has three main purposes: summarizes the content of the e-mail, allows the receiver to determine how important and urgent the message is and allows the user to find the e-mail later on, without having to open it. So give the email a subject/title and keep the subject short and clear. E-mail messages without a subject may not be opened because of a fear of viruses and especially note that it is very easy to forget to type this important information. Short but specific headings are needed, e.g. "Weekly Meeting moved to 2pm this Friday", or "5 Ideas for Joanna's birthday present". Avoid such headings as: "Good News", "Hello", "Team Meeting".

3. Begin with a Greeting

It's important to always open your email with a greeting. Depending on the formality of your relationship, you may want to use their family name as opposed to their given name, i.e. "Dear Mrs. Price,". If the relationship is more casual, you can simply say, "Hi Kelly," If you are contacting a company, not an individual, you may write "To Whom It May Concern".

4. Thank the Recipient

If you are replying to a client's inquiry, you should begin with a line of thanks. For example, if someone has a question about your company, you can say, "Thank you for contacting ABC Company." If someone has replied to one of your emails, be sure to say, "Thank you for your prompt reply." or "Thanks for getting back to me." If you can find any way to thank the reader, then do. It will put him or her at ease, and it will make you appear more courteous.

5. State Your Purpose

If, however, you are initiating the email communication, it may be impossible to include a line of thanks. Instead, start with a clear indication of what your message is about in the first paragraph. For example, "I am writing to enquire about ..." or "I am writing in reference to ..."

It's important to make your purpose clear early on in the email, and then move into the main text of your email. Remember to pay careful attention to grammar, spelling and punctuation, and to avoid run-on sentences by keeping your sentences short and clear.

6. Closing Remarks

Before you end your email, it's polite to thank your reader one more time as well as add some courteous closing remarks. You might start with "Thank you for your patience and cooperation." or "Thank you for your consideration." and then follow up with, "If you have any questions or concerns, don't hesitate to let me know." and "I look forward to hearing from you." Any action that you want the reader to do should be clearly described, using politeness phrases.

7. End with a Closing

The last step is to include an appropriate closing with your name. "Best regards," "Sincerely," and "Thank you," are all professional. It's a good idea to avoid closings such as "Best wishes," or "Cheers," as these are best used in casual, personal emails.

8. Signing off

Include your name at the end of the message. It is most annoying to receive an email which does not include the name of the sender. The problem is that often the email address of the sender does not indicate exactly who it is from, e.g. 1234@gmail.com.

9. Attachments

Make sure you refer, in the main message, to any attachments you are adding and of course make extra sure that you remember to include the attachment(s). As attachments can transmit viruses, try not to use them, unless you are sending complicated documents. If you use an attachment, make sure the file name describes the content, and is not too general.

⊙ Other Useful Information

Please follow these guidelines with all e-mail messages that you send.
- Write a meaningful subject line.
- Keep the message focused and readable.
- Avoid attachments.
- Identify yourself clearly.
- Be kind, don't flame.
- Proofread.
- Don't assume privacy.
- Distinguish between formal and informal situations.
- Respond Promptly.
- Show Respect and Restraint.

⊖Sample

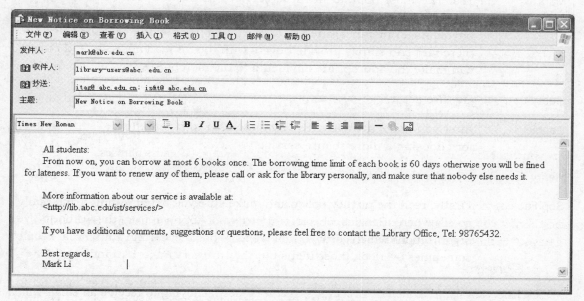

Part 3　Listening and Speaking

⫸⫸ Dialogue: Protecting Buyers' Privacy with Online Payment Services

(Henry was excited to find a long-expected book in an online bookstore, but he hesitated when it required the payment via an account.[1])

Henry	Recently publicized accounts of credit card fraud on the Internet are no doubt alarming to us. It's not always easy to tell who's at the other end of your online transaction when you *hand over* your credit card number.[2]
Mark	Maybe you can use a third-party payment service to make payments for things online a little easier and safer.
Henry	A third-party payment service?
Sophie	Yes. *In brief*, when you use a third-party payment service, you transfer money into an online account and make payments from that account. That way, you never expose your real credit card or bank account information.
Henry	Sounds much safer. Are any of them easily available?

Mark	Yes. The most popular of these services in the U.S. is PayPal owned by eBay, but there are others such as Amazon.com Payments. To take basic payments for instance, such as online auctions or simple Website sales, the merchant can simply provide buyers with their e-mail address, and buyers can make the appropriate payments to the merchant's PayPal account.[3]
Henry	Ok. I've got it.
Sophie	After you make a choice, it's recommended to *be attentive to* some details for choosing a more secure service.
Henry	What're they, please?
Sophie	Firstly, read the privacy policy and make sure you agree with it; If you don't, go elsewhere. Besides, check for a ***stamp*** of approval from the authoritative organizations which certify that a Web site has met certain standards. People sometimes overlook those items but they are very important indeed.[4]
Mark	When using a payment service, please remember, Henry, never respond to e-mail messages from third-party payment services asking you to confirm account details, such as passwords or other personally identifiable information. These e-mail messages could be an identity theft scam, such as ***phishing***.
Henry	Then, how can I confirm or fix my account details more safely?
Sophie	If you need to confirm or update your account information or change your password, I suggest you visit the Website by using your personal bookmark or by typing the URL of the payment service directly into your browser.[5]
Henry	How can I verify further whether a seller providing this payment service is valid?
Mark	You can do it by checking if the seller has been a verified member of the payment service for a few months or more. Some sites also allow you to check the seller's rating. Although these ratings cannot be guaranteed, they can be helpful.
Sophie	*Keep in mind*, Henry, never use your account to transfer money for someone else that you don't know. This might be an advanced fee fraud. And be more careful when you purchase very expensive items, such as jewelry or computers, especially around the holidays and for items that are *sold* in stores.[6]
Henry	Ok. Now I'm much more confident in safe online payments. Thanks a lot for your advice.

▮ Exercises

Work in pairs, and make up a similar conversation by replacing the scenarios with other material in the below.

⊖ Protecting Sellers' Rights in Online Shopping

[1] Want to register an online clothes online store on eBay.

[2] As a seller, how can I ensure that a transaction goes smoothly? And how can I reduce the possibility of *disputes*, *claims*, *chargebacks*, and frauds?

[3] PayPal protects sellers in many ways:

1. Monitor activities and will alert sellers by emails or phones when a suspicious activity occurs.
2. Help sellers to fight unwarranted chargebacks.

   When sellers resolve a buyer's dispute *amicably* through PayPal Dispute Resolution or when they win a PayPal claim, they're covered against a chargeback on that transaction.

[4] Provide clear, detailed descriptions of your item and include photos. Clearly state your return policy. Respond promptly to inquiries.

[5] Before accepting payment:

1. Beware of unusual requests. *Abnormal* requests can be a sign of suspicious activity. A few examples include:
   * *Rush* shipments at any cost.
   * Partial payments from multiple PayPal accounts.
   * Payments not received in full.
2. Be extra cautious with high-priced items, especially if payment is received from one country and sent to another.
3. Know the buyer.
   * Do they have a confirmed address?
   * For e-commerce sites with feedback systems, what is the buyer's score?
   * Do you have any questions? Get answers by emailing or calling the buyer.

[6] Shipping tips:

1. Track packages.

   Packages should be sent with tracking numbers. Once you get a tracking number from the shipping service, promptly send it to the buyer. As an extra *precaution*, you should consider adding delivery confirmation. Finally, make sure to keep proof that the package was received by the buyer.

2. Insure packages.

When sending goods, *opt for* postal insurance to protect yourself in the event that a shipment is lost or the buyer claims it was never delivered.

3. Keep buyers up-to-date about the delivery.

Set expectations *up front*. Once payment is received, give buyers an estimated delivery time. It's better to **overestimate** delivery time than have a package show up late. If a package is not sent in time, tell your buyer about it.

 Words

stamp[stæmp] *n.*印章，戳记	**amicably**[ˈæmikəbli] *adv.*友善地
phish[fiʃ] *v.*网络钓鱼	**abnormal**[æbˈnɔːməl] *adj.*反常的，异常的
dispute[disˈpjuːt] *n.*纠纷，争端	**rush**[rʌʃ] *adj.*紧急的，急迫的
claim[kleim] *n.*索赔，赔款	**precaution**[priˈkɔːʃən] *n.*预防，警惕，防范
chargeback[ˈtʃɑːdʒbæk] *n.*退款	**overestimate**[ˈəuvəˈestimeit] *v.*过高评价，过高估计

 Phrases

hand over	移交，交出	**keep in mind**	记住
in brief	简单地说	**opt for**	选择
be attentive to	注意，留心	**up front**	在前面，预先，先期

≫ Listening Comprehension: Online Shopping

Listen to the passage and the following 3 questions based on it. After you hear a question, there will be a break of 10 seconds. During the break, you will decide which one is the best answer among the four choices marked (A), (B), (C) and (D).

Questions

1. When did the first online store set up mentioned in this article?
2. What is compared to a basket in a conventional store when online shopping?
3. How do consumers find a product of interest in online shopping?

Choices

1. (A) 1990   (B) 1991   (C) 1992   (D) 1994
2. (A) Shopping cart software
   (B) Checkout software
   (C) Permanent online account
   (D) E-mail confirmation
3. (A) By doing a consultation across many different vendors via phone.
   (B) By visiting the Website of the retailer or doing a search via a shopping search engine.

(C) By being concerned about the advertisements on TV.

(D) By collecting the brochures from the physical storefront of the retailer.

 Words

exponentially[ˌekspəu'nenʃəli] v.指数地

 Phrases

by analogy with	由…类推	open up	开始

 Abbreviations

AKA	Also Known As	又名，也称
SSL	Security Socket Layer	安全套接层

≫ Dictation: eBay — Global Buying *Hub*

This passage will be played THREE times. Listen carefully, and fill in the blanks with the words you have heard.

eBay Inc. is an American Internet company that _____1_____ eBay.com, an online _____2_____ and shopping Website in which people and businesses buy and sell goods and services _____3_____ . In addition to its _____4_____ U.S. Website, eBay has established _____5_____ Websites in thirty other countries. Today, eBay is the world's largest online _____6_____ — where *practically* anyone can sell practically anything at any time. It's an idea that BusinessWeek oncecalled "*nothing less than* a virtual, self-regulating global _____7_____ ."

eBay was _____8_____ in 1995, when Pierre Omidyar, a French-born Iranian computer programmer, wrote the code for an auction Website that he *ran* from his home computer. According to company *lore*, Mr. Omidyar was _____9_____ to create an online service for _____10_____ , person-to-person sales transactions by his wife's interest in _____11_____ Pez candy *dispensers*. The first _____12_____ sold on eBay was a broken laser pointer for $14.83. *Astonished*, Omidyar contacted the winning *bidder* to ask if he _____13_____ that the laser pointer was broken. In his responding email, the buyer explained: "I'm a collector of broken laser pointers."

The company has three business _____14_____: Marketplaces, Payments and Communications. Its Marketplaces segment enables online commerce through a variety of platforms, including the traditional eBay.com platform and eBay's other online platforms, such as its ___15___ Websites, as well as Half.com, Rent.com, Shopping.com and StubHub. eBay's Payments segment, which consists of PayPal, enables ___16___ and businesses to send and receive payments online. Its Communications segment, which consists of Skype, enables VoIP communications between Skype users and provides low-cost ___17___ to traditional *fixed-line* and mobile telephones.

With a ___18___ in 39 markets, including the U.S., and approximately 84 million ___19___ users worldwide, eBay has changed the ___20___ of Internet commerce. In 2007, the total value of sold items on eBay's trading platforms was nearly $60 billion. This means that eBay users worldwide trade more than $1,900 worth of goods on the site every second.

 Words

hub[hʌb] *n.*（兴趣、活动等的）中心	**dispenser**[dis'pensə] *n.* 自动售货机
practically['præktikəli] *adv.* 几乎，差不多	**astonish**[əs'tɔniʃ] *v.* 使惊讶
run[rʌn] *v.*（工作等）进行	**bidder**['bidə] *n.* 出价人，投标人
lore[lɔ:] *n.* 口头传说	

 Phrases

nothing less than	恰恰是	**fixed-line**	固定电话

Computer Security

Unit
10

Part 1   Reading and Translating

Section A Computer Security

Computer security is a branch of technology known as information security as applied to computers. The objective of computer security varies and can include protection of information from theft or **corruption**, or the preservation of availability, as defined in the security policy.

*Computer security **imposes** requirements on computers that are different from most system requirements because they often take the form of **constraints** on what computers are not supposed to do.* This makes computer security particularly challenging because it is hard enough just to make computer programs do everything they are designed to do correctly. *Furthermore, negative requirements are **deceptively** complicated to satisfy and require **exhaustive** testing to verify, which is impractical for most computer programs.* Computer security provides a technical strategy to convert negative requirements to positive **enforceable** rules. For this reason, computer security is often more technical and mathematical than some other computer science fields.

Typical approaches to improving computer security (in approximate order of strength) can include the following:

- Physically limit access to computers to only those who will not compromise security.
- Hardware mechanisms that impose rules on computer programs, thus avoiding depending on computer programs for computer security.
- Operating system mechanisms that impose rules on programs to avoid trusting computer programs.
- Programming strategies to make computer programs dependable and resist *subversion*.

Applications in Aviation

Computer security is critical in almost any technology-driven industry operating on computer systems. The issues of computer based systems and addressing their countless vulnerabilities are an integral part of maintaining an operational industry.

The aviation industry is especially important when analyzing computer security because the involved risks include expensive equipment and *cargo*, transportation infrastructure, and human life. Security can be compromised by hardware and software *malpractice*, human error, and faulty operating environments. Threats that exploit computer vulnerabilities can stem from *sabotage*, *espionage*, industrial competition, terrorist attack, mechanical malfunction, and human error.

The consequences of a successful *deliberate* or *inadvertent* misuse of a computer system in the aviation industry range from loss of *confidentiality* to loss of system integrity, which may lead to more serious *concerns* such as data theft or loss, network and air traffic control *outages*, which in turn can lead to airport closures, loss of aircraft, loss of passenger life. Military systems that control *munitions* can pose an even greater risk.

A proper attack does not need to be very high tech or well funded: for a power outage at an airport alone can cause *repercussions* worldwide. One of the easiest and, arguably, the most difficult to trace security vulnerabilities is achievable by transmitting unauthorized communications over specific radio frequencies. These transmissions may *spoof* air traffic controllers or simply *disrupt* communications altogether. These incidents are very common, having altered flight courses of commercial aircraft and caused panic and confusion in the past. Controlling aircraft over oceans is especially dangerous because radar *surveillance* only extends 175 to 225 miles offshore. Beyond the radar's sight controllers must rely on periodic radio communications with a third party.

Lightning, power *fluctuations*, *surges*, *brown-outs*, *blown fuses*, and various other power outages instantly disable all computer systems, since they are dependent on electrical source. *Other accidental and intentional faults have caused significant disruption of safety critical systems* *throughout the last few decades and dependence on reliable communication and electrical power only **jeopardizes** computer safety.*

Notable System Accidents

In 1983, Korean Airlines Flight 007, a Boeing 747 was shot down by Soviet Su-15 *jets* after a navigation computer malfunction caused the aircraft to *steer* 185 miles off *course* into Soviet Union airspace. All 269 passengers were killed.

In 1994, over a hundred intrusions were made by unidentified hackers into the Rome Laboratory, the US Air Force's main command and research facility. Using *Trojan horse* viruses, hackers were able to obtain unrestricted access to Rome's networking systems and remove traces of their activities. The intruders were able to obtain classified files, such as air tasking order systems data and furthermore able to penetrate connected networks of National Aeronautics and Space Administration's Goddard Space Flight Center, Wright-Patterson Air Force Base, some Defense contractors, and other private sector organizations, by *posing as* a trusted Rome center user.

Electromagnetic interference is another threat to computer safety. In 1989, a United States Air Force F-16 jet accidentally dropped a 230 kg bomb in West Georgia after unspecified interference caused the jet's computers to release it.

A similar telecommunications accident also happened in 1994, when two UH-60 Blackhawk helicopters were destroyed by F-15 aircraft in Iraq because the **IFF** system's encryption system malfunctioned.

Terminology

The following terms used in engineering secure systems are explained below.

- **Firewalls** can either be hardware devices or software programs. They provide some protection from online intrusion, by restricting the network traffic which can pass through them, based on a set of system administrator defined rules. However, since they allow some applications (e.g. Web browsers) to connect to the Internet, they don't protect against some unpatched vulnerabilities in these applications.
- **Automated *theorem* proving** and other verification tools can enable critical algorithms and code used in secure systems to be mathematically proven to meet their specifications.
- Simple **microkernels** can be developed so that we can be sure they don't contain any bugs: e.g. EROS and Coyotos. A bigger OS, capable of providing a standard API like POSIX, can be built on a secure microkernel using small API servers running as normal programs. If one of these API servers has a bug, the kernel and the other servers are not affected: e.g. Hurd or Minix 3.

- *Cryptographic* **techniques** can be used to defend data in transit between systems, reducing the probability of intercepting or modifying those data being exchanged between systems. Cryptographic techniques involve transforming information, *scrambling* it so it becomes unreadable during transmission. The intended recipient can unscramble the message, but *eavesdroppers* cannot.

- **Secure cryptoprocessors** can be used to leverage physical security techniques into protecting the security of the computer system.

- **Chain of trust techniques** can be used to attempt to ensure that all software loaded has been certified as *authentic* by the system's designers.

- **Access authorization** restricts access to a computer from group of users by authentication systems. These systems can protect either the whole computer – such as through an interactive logon screen or individual services, such as an **FTP** server. There are many methods for identifying and authenticating users, such as passwords, identification cards, and, more recently, smart cards and *biometric* systems.

- *Mandatory* **access control** can be used to ensure that *privileged* access is withdrawn when privileges are *revoked*. For example, deleting a user account should also stop any processes that are running with that user's privileges.

- **Backups** are another copy of all the important computer files kept in another location for information securing. These files are kept on hard disks, **CD-Rs**, **CD-RW**s, and tapes. It also involves using one of the file hosting services that backs up files over the Internet for both business and individuals.

- **Anti-virus software** consists of computer programs that attempt to identify, *thwart* and eliminate computer viruses and other malicious software (malware).

- **Intrusion-detection systems** can scan a network for users that are on the network but who should not be there or are doing things that they should not be doing, for example trying a lot of passwords to gain access to the network.

- **Social engineering awareness** keeps employees aware of the dangers of social engineering and/or having a policy *in place* to prevent social engineering can reduce successful *breaches* of the network and servers.

- *Honey pots* are computers that are either intentionally or unintentionally left vulnerable to attack by *crackers*. They can be used to catch crackers or fix vulnerabilities.

 Words

corruption[kə'rʌpʃən] *n.*变坏	**deceptively**[di'septivli] *adv.*欺骗地，虚伪地
impose[im'pəuz] *v.*把…强加给，强派	**exhaustive**[ig'zɔːstiv] *adj.*无遗漏的，详尽的
constraint[kən'streint] *n.*约束，强制	**enforceable**[in'fɔːsəbl] *adj.*可实施的，可执行的

subversion[sʌb'və:ʃən] n.颠覆，破坏

cargo['kɑ:gəu] n.货物

malpractice['mæl'præktis] n.玩忽职守，业务技术事故

sabotage['sæbətɑ:ʒ] n.蓄意破坏

espionage['espiənidʒ] n.间谍活动

deliberate[di'libəreit] adj.故意的，蓄意的

inadvertent[,inəd'və:tənt] adj.疏忽造成的

confidentiality[,kɔnfi,denʃi'æliti] n.机密性

concern[kən'sə:n] n.担心，忧虑

outage['autidʒ] n.运转中断，停用

munitions[mju(:)'niʃəns] n.军需品，战争物资

repercussion[,ri:pə(:)'kʌʃən] n.反响

spoof[spu:f] v.哄骗，戏弄

disrupt[dis'rʌpt] v.干扰，扰乱，使中断

surveillance[sə'veiləns] n.监视

fluctuation[,flʌktju'eiʃən] n.波动，起伏

surge[sə:dʒ] n.电涌

brown-out['braunaut] n.部分灯火管制

blow[bləu] v.烧断，熔断

fuse[fju:z] n.保险丝

jeopardize['dʒepədaiz] v.危害

jet[dʒet] n.喷气机

steer[stiə] v.驾驶

course[kɔ:s] n.航道

theorem['θiərəm] n.定理，法则

cryptographic[,kriptə'græfik] adj.密码的，暗号的

scramble['skræmbl] v.搅乱，使混杂

eavesdropper['i:vz,drɔpə] n.偷听者

authentic[ɔ:'θentik] adj.可信的，可靠的

biometric[,baiəu'metrik] adj.生物鉴别法的

mandatory['mændətəri] n.强制的，托管的

privilege['privilidʒ] v.给与…特权

revoke[ri'vəuk] v.取消，废止

thwart[θwɔ:t] v.阻止…的发生

breach[bri:tʃ] n.破坏

cracker['krækə] n.骇客

 Phrases

Trojan horse	特洛伊木马	in place	在适当的地位
pose as	假装，冒充	honey pot	贮蜜罐

 Abbreviations

IFF	Identification Friend or Foe	敌我识别
FTP	File Transfer Protocol	文件传输协议
CD-R	Compact Disc-Recordable	可记录光盘
CD-RW	Compact Disc ReWritable	可擦写光盘

 Complex Sentences

1. **Original:** Computer security imposes requirements on computers that are different from most system requirements because they often take the form of constraints on what computers are not supposed to do.

Translation: 计算机安全会将需求强加到计算机上，这些需求不同于大多数系统需求，因为这些需求通常表现为强制的形式，而这些强制是在计算机不允许做的事情方面的强制。

◼◼◼ **Exercises**

I. Read the following statements carefully, and decide whether they are true (T) or false (F) according to the text.

___1. Computer security is a branch of technology arranged under information security category.

___2. Physically limiting access to computers is approximately one of the strongest approaches to improve computer security.

___3. Computer programming strategies are often considered as the most dependent approach for computer security among others.

___4. In aviation, transmitting unauthorized communications over specific radio frequencies may spoof air traffic controllers or simply disrupt communications altogether.

___5. The issues of computer security are an indispensable part of maintaining an operational industry.

II. Choose the best answer to each of the following questions.

1. What is the primary difference of computer security from some other computer science fields?
   (A) Its varied objectives   (B) Its positive enforceable rules
   (C) Its negative requirements   (D) Its impractical for most computer programs

2. Which of the following is WRONG about the terminology used in engineering secure systems?
   (A) Firewalls can restrict the network traffic which can pass through them, based on a set of system administrator defined rules.
   (B) Microkernels can be used in conjunction with small API servers running as normal programs to build a big OS.
   (C) Cryptography can transform data into something unreadable and unrecoverable.
   (D) Intrusion-detection systems can scan suspicious accesses or manipulations on a network.

3. Which of the following is NOT argued in the instances of computer security accidents in aviation?
   (A) Human error   (B) Computer malfunction
   (C) Hackers intruding   (D) Electromagnetic interference

III. Translating.

1. Furthermore, negative requirements are deceptively complicated to satisfy and require exhaustive testing to verify, which is impractical for most computer programs.

2. Other accidental and intentional faults have caused significant disruption of safety critical systems throughout the last few decades and dependence on reliable communication and electrical power only jeopardizes computer safety.

⫸ Section B   Computer Virus

A computer virus is a computer program that can copy itself and infect a computer without permission or knowledge of the user. The term "virus" is also commonly used, *albeit* erroneously, to refer to many different types of *malware* and *adware* programs. The original virus may modify the copies, or the copies may modify themselves, as occurs in a ***metamorphic*** virus. A virus can only spread from one computer to another when its host is taken to the uninfected computer, for instance by a user sending it over a network or the Internet, or by carrying it on a removable medium such as a floppy disk, CD, or USB drive. Meanwhile viruses can spread to other computers by infecting files on a network file system or a file system that is accessed by another computer.

Most personal computers are now connected to the Internet and to local area networks, facilitating the spread of malicious code. Today's viruses may also take advantage of network services such as the World Wide Web, e-mail, Instant Messaging and file sharing systems to spread. Furthermore, some sources use an alternative terminology in which a virus is any form of self-replicating malware.

Some malware is programmed to damage the computer by damaging programs, deleting files, or reformatting the hard disk. *Other malware programs are not designed to do any damage, but simply replicate themselves and perhaps make their presence known by presenting text, video, or audio messages.* Even these less ***sinister*** malware programs can create problems for the computer user. They typically *take up* computer memory used by legitimate programs. As a result, they often cause ***erratic*** behavior and can result in system crashes. In addition, much malware is bug-ridden, and these bugs may lead to system crashes and data loss. Many ***CID*** programs are programs that have been downloaded by users and *pop up every so often*. This results in slowing down of the computer, but it is also very difficult to discover and solve the problem.

⊜ Infection Strategies

In order to replicate itself, a virus must be permitted to execute code and write to memory. For this reason, many viruses attach themselves to executable files that may be part of legitimate programs. If a user tries to start an infected program, the virus' code may be executed first. Viruses can be divided into two types, on the basis of their behavior when they are executed. Nonresident viruses immediately search for other ***hosts*** that can be infected, infect these targets, and finally transfer control to the application program they infected. Resident viruses do not search for hosts when they are started. Instead, a resident virus loads itself into memory on execution and transfers control to the host program. The virus stays active in the background and infects new hosts when those files are accessed by other programs or the operating system itself.

1. Nonresident Viruses

Nonresident viruses can be thought of as consisting of a finder module and a replication module. The finder module is responsible for finding new files to infect. For each new executable file the finder module encounters, it calls the replication module to infect that file.

2. Resident Viruses

Resident viruses contain a replication module that is similar to the one that is employed by nonresident viruses. However, this module is not called by a finder module. Instead, *the virus loads the replication module into memory when it is executed and ensures that this module is executed each time the operating system is called to perform a certain operation.* For example, the replication module can be called each time the operating system executes a file. In this case, the virus infects every suitable program that is executed on the computer.

Resident viruses are sometimes subdivided into a category of fast infectors and a category of slow infectors. Fast infectors are designed to infect as many files as possible. For instance, a fast infector can infect every potential host file that is accessed. This poses a special problem to anti-virus software, since a virus scanner will access every potential host file on a computer when it performs a system-wide scan. If the virus scanner fails to notice that such a virus is present in memory, the virus can "*piggy-back*" on the virus scanner and in this way infect all files that are scanned. Fast infectors rely on their fast infection rate to spread. The disadvantage of this method is that infecting many files may make detection more likely, because the virus may slow down a computer or perform many suspicious actions that can be noticed by anti-virus software. Slow infectors, on the other hand, are designed to infect hosts infrequently. For instance, some slow infectors only infect files when they are copied. Slow infectors are designed to avoid detection by limiting their actions: they are less likely to slow down a computer noticeably, and will at most infrequently trigger anti-virus software that detects suspicious behavior by programs. The slow infector approach does not seem very successful, however.

⊘ Vulnerability and *Countermeasures*

1. The Vulnerability of Operating Systems to Viruses

*Just as **genetic** diversity in a **population** decreases the chance of a single disease wiping out a population, the diversity of software systems on a network similarly limits the destructive potential of viruses.*

This became a particular concern in the 1990s, when Microsoft gained market dominance in desktop operating systems and office suites. The users of Microsoft software (especially networking software such as Microsoft Outlook and Internet Explorer) are especially vulnerable to the spread of viruses. Microsoft software is **targeted** by virus writers due to their desktop dominance, and is often criticized for including many errors and holes for virus writers to exploit. Integrated and non-integrated Microsoft applications (such as Microsoft Office) and applications with scripting languages with access to the file system (for example Visual Basic Script (**VBS**), and applications with networking features) are also particularly vulnerable.

Although Windows is by far the most popular operating system for virus writers, some viruses also exist on other platforms. Any operating system that allows third-party programs to run can theoretically run viruses. Some operating systems are less secure than others. Unix-based OS's (and **NTFS**-aware applications on Windows NT based platforms) only allow their users to run executables within their protected space in their own directories.

As of 2006, there are relatively few security exploits targeting Mac OS X (with a Unix-based file system and kernel). The number of viruses for the older Apple operating systems, known as Mac OS Classic, varies greatly from source to source, with Apple stating that there are only four known viruses, and independent sources stating there are as many as 63 viruses. It is safe to say that Macs are less likely to be targeted because of low market share and thus a Mac-specific virus could only infect a small proportion of computers (making the effort less **desirable**). Virus vulnerability between Macs and Windows is a chief selling point, one that Apple uses in their Get a Mac advertising.

Windows and Unix have similar scripting abilities, but while Unix natively blocks normal users from having access to make changes to the operating system environment, older copies of Windows such as Windows 95 and 98 do not. *In 1997, when a virus for Linux was released – known as "Bliss" – leading antivirus vendors issued warnings that Unix-like systems could fall prey to viruses just like Windows.* The Bliss virus may be considered as a characteristic virus – as opposed to worms – on Unix systems. Bliss requires that the user run it explicitly (so it is a Trojan), and it can only infect programs that the user has the access to modify. Unlike Windows users, most Unix users do not log in as an administrator user except to install or configure software; as a result, even if a user ran the virus, it could not harm their operating system. The Bliss virus never became widespread, and remains chiefly a research **curiosity**. Its creator later posted the source code to **Usenet**, allowing researchers to see how it worked.

2. The Role of Software Development

Because software is often designed with security features to prevent unauthorized use of system resources, many viruses must exploit software bugs in a system or application to spread. Software development strategies that produce large numbers of bugs will generally also produce potential exploits.

3. Anti-virus Software

Many users install anti-virus software that can detect and eliminate known viruses after the computer downloads or runs the executable files. There are two common methods that an anti-virus software application uses to detect viruses. The first, and by far the most common method of virus detection is using a list of virus **signature** definitions. This works by examining the content of the computer's memory (its RAM, and *boot sectors*) and the files stored on fixed or removable drives (hard drives, floppy drives), and comparing those files against a database of known virus "signatures". The disadvantage of this detection method is that users are only protected from viruses that **predate** their last virus definition update. The second method is to use a **heuristic** algorithm to find viruses based on common behaviors. This method has the ability to detect viruses that anti-virus security firms have yet to create a signature for.

Some anti-virus programs are able to scan opened files in addition to sent and received e-mails "*on the fly*" in a similar manner. This practice is known as "*on-access scanning*." Anti-virus software does not change the underlying capability of host software to transmit viruses. Users must update their software regularly to patch security holes. Anti-virus software also needs to be regularly updated in order to prevent the latest threats.

4. Other Preventive Measures

One may also minimize the damage done by viruses by making regular backups of data (and the operating systems) on different media, that are either kept unconnected to the system (most of the time), read-only or not accessible for other reasons, such as using different file systems. In this way, if data is lost through a virus, one can start again using the backup (which should preferably be recent). A notable exception to this rule is the Gammima virus, which **propagates** via infected removable media (specifically *flash drives*). If a backup session on optical media like CD and DVD is closed, it becomes read-only and can no longer be affected by a virus (so long as a virus or infected file was not copied onto the CD/DVD). Likewise, an operating system on a bootable can be used to start the computer if the installed operating systems become unusable. Another method is to use different operating systems on different file systems. A virus is not likely to affect both. Data backups can also be put on different file systems. For example, Linux requires specific software to write to NTFS partitions, so if one does not install such software and uses a separate installation of MS Windows to make the backups on an NTFS partition, the backup should remain safe from any Linux viruses. Likewise, MS Windows can not read file systems like **ext3**, so if one normally uses MS Windows, the backups can be made on an ext3 partition using a Linux installation.

Words

albeit[ɔ:l'bi:it] *conj.*虽然	**target**['tɑ:git] *v.*把…作为目标
metamorphic[,metə'mɔ:fik] *adj.*变形的，变质的	**desirable**[di'zaiərəbl] *adj.*值得做的，值得要的，有利的
sinister['sinistə] *adj.*险恶的	**prey**[prei] *n.*猎物，牺牲品
erratic[i'rætik] *adj.*奇怪的	**curiosity**[,kjuəri'ɔsiti] *n.*好奇心
host[həust] *n.*主机	**signature**['signitʃə] *n.*特征代码
countermeasure['kauntə,meʒə] *n.*对策，反措施	**predate**['pri:'deit] *v.*先于，时间上先于…
genetic[dʒi'netik] *adj.*遗传学的，基因的	**heuristic**[hjuə'ristik] *adj.*启发式的
population[,pɔpju'leiʃən] *n.*种群	**propagate**['prɔpəgeit] *v.*繁殖，传播

 Phrases

malware	恶意软件	**wipe out**	消灭
adware	广告软件	**fall to**	着手开始工作
take up	占空间	**boot sector**	引导扇区
CID	一种浏览器病毒	**on the fly**	（计算机）运行中，勿忙地，忙碌地
pop up	弹出	**on-access scanning**	存取时扫描
every so often	时常，不时	**flash drive**	闪存
piggy-back	背负式装运		

 Abbreviations

VBS	Visual Basic Script	VB脚本
NTFS	New Technology File System	Windows NT以上版本支持的一种文件系统
Usenet	Uses Network	新闻讨论组
Ext3	Third extended file system	一种日志式文件系统

 Complex Sentences

1. **Original:** Other malware programs are not designed to do any damage, but simply replicate themselves and perhaps make their presence known by presenting text, video, or audio messages.

   Translation: 其他恶意软件程序设计的目的不是造成任何破坏，而是简单地自我复制并可能通过展示文本、视频或音频消息让人们知道它们的存在。

2. **Original:** Just as genetic diversity in a population decreases the chance of a single disease wiping out a population, the diversity of software systems on a network similarly limits the destructive potential of viruses.

Translation: 正像在种群中的遗传多样性降低了某种单一疾病灭绝种群的可能性，网络中软件系统的多样性也类似地限制了病毒的破坏性潜力。

 **Exercises**

I. Read the following statements carefully, and decide whether they are true (T) or false (F) according to the text.

___·1. Windows is by far the most popular operating system for virus writers due to their market dominance.

___2. Bugs in software generally means potential exploits for the spread of viruses.

___3. The method of using a heuristic algorithm to find viruses can detect viruses which haven't yet been defined in the current database of known virus "signatures".

___4. The backups on CD/DVD are always safe because the media becomes read-only and can no longer be affected by a virus.

___5. Anti-virus software can enhance the underlying capability of host software to prevent viruses transmitting.

II. Choose the best answer to each of the following questions.

1. Which of the following is RIGHT about the infection strategies of viruses?

(A) Viruses may take advantage of present network services to facilitate their spread.

(B) All original viruses must modify the copies in metamorphic viruses for spread.

(C) The success of viruses all relies on their fast infection rate to spread.

(D) All viruses pretend that they are executable files in order to replicate themselves.

2. Which of the following is NOT an effective countermeasure to viruses?

(A) Updating operating system regularly to patch security holes

(B) Installing and regularly updating anti-virus software

(C) Making regular backups of data on different file systems

(D) Using different operating systems on the same file system

3. Why does the author give the example of "Bliss"?

(A) To argue that the Bliss virus has characteristic of viruses as opposed to worms.

(B) To argue that the Bliss virus could be harmless even if users ran it on Unix.

(C) To argue that other operating systems may run viruses, not just Windows.

(D) To argue that the Bliss virus only remains chiefly a research curiosity.

III. Translating.

1. The virus loads the replication module into memory when it is executed and ensures that this module is executed each time the operating system is called to perform a certain operation.

2. In 1997, when a virus for Linux was released – known as "Bliss" – leading antivirus vendors issued warnings that Unix-like systems could fall prey to viruses just like Windows.

Part 2   **Simulated Writing: Business Letter**

⬤ Introduction

A good business letter is brief, straightforward, and polite. If possible, it should be limited to one single-spaced typewritten page. Because it is so brief, a business letter is often judged on detailed, but important, things: format, grammar, punctuation, openings and closings. A business letter is not the place to try out fancy fonts or experimental writing styles.

⬤ Styles of Business Letters

There are two main styles of business letters.
- **Full block style:** Align all elements on the left margin.
- **Modified block style:** Down the middle of the page, align the return address, date, closing, signature, and typed name; align other elements on the left page margin.

⬤ Parts of a Business Letter

1. Sender's Address

Including the address of the sender is optional. If you choose to include it, place the address on the top of the page. Do not write the sender's name or title, as it is included in the letter's closing. Include only the street address, city and zip code. Another option is to include the sender's address directly after the closing signature.

2. Date

The date line is used to indicate the date the letter was written. When writing to companies within the United States, use the American date format. (The United States-based convention for formatting a date places the month before the day. For example: June 11, 2001.) The date line leaves two blank lines after the sender's address block. Depending which format you are using for your letter, either left justify the date or center it horizontally.

3. Inside Address

The inside address is the recipient's address. It is always best to write to a specific individual at the firm to which you are writing. If you do not have the person's name, do some research by calling the company or speaking with employees from the company. Include a personal title such as Ms., Mrs., Mr., or Dr. Follow a woman's preference in being addressed as Miss, Mrs., or Ms. If you are unsure of a woman's preference in being addressed, use Ms. If there is a possibility that the person to whom you are writing is a Dr. or has some other title, use that title. Usually, people will not mind being addressed by a higher title than they actually possess. To write the address, use the U.S. post office format. For international addresses, type the name of the country in all-capital letters on the last line. The inside address block leaves two blank lines after the date. It should be left justified, no matter which format you are using.

4. Salutation

Use the same name as the inside address, including the personal title. If you know the person and typically address them by their first name, it is acceptable to use only the first name in the salutation (for example: Dear Lucy:). In all other cases, however, use the personal title and full name followed by a colon. Leave one line blank after the salutation.

If you don't know the reader's gender, use a nonsexist salutation, such as "To Whom it May Concern." It is also acceptable to use the full name in a salutation if you cannot determine gender. For example, you might write Dear Chris Harmon: if you were unsure of Chris's gender.

5. Body

For block and modified block formats, single space and left justify each paragraph within the body of the letter. Leave a blank line between each paragraph. When writing a business letter, be careful to remember that conciseness is very important. In the first paragraph, consider a friendly opening and then a statement of the main point. The next paragraph should begin justifying the importance of the main point. In the next few paragraphs, continue justification with background information and supporting details. The closing paragraph should restate the purpose of the letter and, in some cases, request some type of action.

6. Closing

The closing begins at the same horizontal point as your date and one line after the last body paragraph. Capitalize the first word only (for example: Thank you) and leave four lines between the closing and the sender's name for a signature. If a colon follows the salutation, a comma should follow the closing; otherwise, there is no punctuation after the closing. Leave two lines of space after your last body paragraph, then use a conventional closing, followed by a comma (i.e., Sincerely, Sincerely Yours, Respectfully).

7. Enclosures

If you have enclosed any documents along with the letter, such as a resume, you should indicate this simply by typing Enclosures one line below the closing. As an option, you may list the name of each document you are including in the envelope. For instance, if you have included many documents and need to ensure that the recipient is aware of each document, it may be a good idea to list the names.

8. Signature

Your signature should appear below your closing. Unless you have established a personal relationship with the person you are writing to, use both your first and last name.

9. Name and Position

Four lines after the closing, type your full name. Do not include a title (Mr. or Mrs.). If you are writing on behalf of an organization, type your title on the next line.

Abbreviations Used in Letter Writing

The following abbreviations are widely used in letters:
- **ASAP** = as soon as possible
- **Cc** = carbon copy (when you send a copy of a letter to more than one person, you use this abbreviation to let them know)
- **Enc. or Enclosure** = enclosure (when you include other papers with your letter)
- **PP** = per procurationem (A Latin phrase meaning that you are signing the letter on somebody else's behalf; if they are not there to sign it themselves, etc)
- **PS** = postscript (when you want to add something after you've finished and signed it)
- **PTO (informal)** = please turn over (to make sure that the other person knows the letter continues on the other side of the page)
- **RSVP** = please reply

Sample

Department of Computer Science and Technology
Beihang University
Beijing, 100191
P.R. China

March 15, 2004

Prof. Joseph Wood
Department of Computer
New York University
New York
USA

Dear Prof. Wood,

It is a great pleasure that I invite you to the 11th Annual Computer Society Conference. This year's conference will be held at Beijing from September 23 to September 27.

We are offering a valuable program with industry-wide applications, speakers who are recognized experts in their field and topics with many implications for the future. Ample time is scheduled for discussion periods. In addition, tours to two large computer companies have been arranged.

Enclosed please find information on accommodations, transportation and registration. If you have any questions, please call the session coordinator, Julie Han at 8610-12345678.

Yours Sincerely
(Signature)
Chen Jun
Chairman
Department of Computer Science

Part 3 Listening and Speaking

⫸ Dialogue: Using Anti-virus Software

*(Sophie's computer is unfortunately infected by a malicious email attachment. Now she is trying to find some measures to **cure** and protect her computer from virus infections effectively.)*

Mark	I think anti-virus software is one of the main defenses against online problems.
Henry	Yes. I agree. Anti-virus software covers the main lines of attack. It scans incoming emails for attached viruses, monitors files as they are opened osr created to make sure they are not infected and performs periodic scans of every file on the computer.[1]
Sophie	Could you suggest any reputable anti-virus software for me to use?
Mark	There are many suppliers of anti-virus software.Companies that make commercial anti-virus software include F-Secure, Kaspersky, McAfee, Microsoft, Symantec, Trend Micro, and so on.

Henry	On the other hand, there are also some free anti-virus software you can choose, such as Grisoft AVG Anti-Virus, AntiVir, ALWIL Avast and ClamWin.
Sophie	How to choose a proper one among so many products?
Henry	In most cases, these "free" products are *scaled-back* versions of commercial products to which the software manufacturer hopes you will, one day, upgrade. And they are often limited or no technical support and reduced functionality.
Mark	So, unless getting free software is critical, it is preferable to buy a fully-supported commercial product.[2]
Henry	For personal and home office uses there are a number of basic choices that you can make to decide which anti-virus software to buy. For example, most anti-virus software companies sell a standalone program as well as security *suite* packages.
Sophie	What are the differences between standalone anti-virus and security suite?
Henry	A standalone program only scans for viruses while security suite packages include other protective software such as a firewall, spam filtering, anti-*spyware* and so on.
Mark	So the advantages of a suite include covering all the bases, sharing a single user interface and being easier and cheaper than buying each individual program separately.[3] For beginners, it is probably best to buy a suite.
Sophie	So, does it mean that my computer would be safe enough as long as anti-virus software is installed?
Henry	*Not yet.* Please remember, Sophie, anti-virus software will not protect you against any attack at any time. For example, it can't protect you against programs that you choose to install and that may contain unwanted features, spam, any kind of fraud or criminal activity online or a hacker trying to *break into* your computer over the Internet.[4]
Mark	It is less effective if it is not kept up-to-date with the latest virus signatures.

Sophie	A virus signature?
Mark	Yes. A virus signature is like a criminal's *mug shot*. Each time a new virus is released, security firms analyze it and create a new signature that lets anti-virus software block the new virus.
Sophie	Ok, I've got it. Thank you very much!
Henry & Mark	Not at all. Good luck, Sophie!

Exercises

Work in pairs, and make up a similar conversation by replacing the scenarios with the material below.

Using a Firewall

[1] A firewall acts as a barrier between the public Internet and your private computer or network as the first line of defense. It protects you against a number of different online threats:
- Hackers breaking into your computer.
- Some viruses, called "**worms**," that spread from computer to computer over the Internet.
- Some firewalls block outgoing traffic that might originate from a virus infection.

[2] Desktop firewall (also sometimes known as "software firewall") and hardware firewall
- Desktop firewall: A desktop firewall is installed on each computer that is connected to the Internet and monitors (and blocks, where necessary) Internet traffic. A desktop firewall on each computer is your first priority.
- Hardware firewalls are often built into broadband Internet routers. If several computers share an Internet connection, a hardware firewall will protect all of them. Most router manufacturers offer devices with firewalls.
- You can have hardware and desktop firewalls and having both may give a small margin of extra security.

[3] Windows Firewall and Commercial firewalls
- Windows Firewall is a free basic desktop firewall and is included with Windows Vista and Windows XP (with the latest updates).
- Commercial firewalls operate in the same way as Windows Firewall but generally give you extra protection, more control over how the firewall works and more information

about how to configure it. Commercial desktop firewalls often integrate well with other security products like virus scanners.

[4] A firewall isn't sufficient on its own to guarantee security. You also need to take the other protective steps. However, a firewall provides limited or no protection at all:

- If you give permission for other computers to connect to yours.
- Against most viruses, spam or spyware installations.
- If you or a virus has created a back door through the firewall.
- If a hacker has the password for the firewall.
- Against malicious traffic that does not travel through it.

Words

cure[kjuə] v.改正，消除，治疗 suite[swi:t] n.套装，套件	worm[wə:mz] n.计算机网络 "蠕虫"

Phrases

scaled-back	按比例缩小	mug shot	大头照，面部特写
spyware	间谍软件	break into	闯进
not yet	还没有		

Listening Comprehension: Hacker and Cracker

Listen to the passage and the following 3 questions based on it. After you hear a question, there will be a break of 10 seconds. During the break, you will decide which one is the best answer among the four choices marked (A), (B), (C) and (D).

Questions

1. Which term is a positive title originally?
2. Why has the popular definition of hacker been changed?
3. Which of the following behavior(s) is (are) illegitimate?

Choices

1. (A) Hacker   (B) Cracker   (C) Phishers   (D) None of above
2. (A) Due to sensationalized depictions from industrial experts
   (B) Due to exaggerated depictions from social critics
   (C) Due to sensationalized depictions in modern media
   (D) Due to exaggerated depictions from hackers themselves

3. (A) Tricking others' credit card numbers via computer

   (B) Tampering with files in others' computers without permission

   (C) Breaking into networks with malicious intent

   (D) All of above

 Words

amateur['æmətə] *n.*业余爱好者	**tamper**['tæmpə] *v.*篡改
recondite[ri'kɔndait] *adj.*深奥的	**legitimate**[l i'dʒitimit] *adj.*合法的
tinker['tiŋkə] *v.*笨手笨脚地做事，修补	**resent**[ri'zent] *v.*憎恶，怨恨
sensationalize[sen'seiʃ ən,laiz] *v.*加以渲染，使耸人听闻	**phisher**['fiʃ ə] *n.*网络钓鱼者
snoop[snu:p] *v.*窥探，偷窃	**trick**[trik] *v.*哄骗，欺诈

 Phrases

take into account	考虑

>>> Dictation: Trojan Horses

   This passage will be played THREE times. Listen carefully, and fill in the blanks with the words you have heard.

   The name "Trojan horse" comes from a legend told in the *Iliad* about the *siege* of the city of Troy by the Greeks. Legend has it that the Greeks, unable to penetrate the city's _____1_____ , got the idea to give up the siege and instead give the city a giant wooden horse as a gift _____2_____ . The Trojans (the people of the city of Troy) accepted this seemingly _____3_____ gift and brought it within the city walls. However, the horse was _____4_____ with soldiers, who came out at *nightfall*, while the town _____5_____ , to open the city gates so that the ____6____ of the army could enter.

   Thus, in the world of computing, a Trojan horse is any program that _____7_____ the user to run it, but *conceals* a harmful or malicious *payload*, and usually opens up _____8_____ to the computer running it by opening a _____9_____ . For this reason, it is sometimes called a Trojan by analogy to the citizens of Troy. The payload may take effect _____10_____ and can lead to many undesirable effects, such as deleting all the user's files, or more commonly it may _____11_____ further harmful software into the user's system to serve the creator's longer-term _____12_____ . Trojan horses known as *droppers* are used to *start off* a worm _____13_____ , by injecting the worm into users' local networks.

Since Trojan horses have a _____14_____ of forms, there is no single method to delete them. The simplest responses involve clearing the _____15_____ Internet file and deleting it _____16_____. Normally, antivirus software is able to detect and remove the Trojan _____17_____. If the antivirus cannot find it, booting the computer from _____18_____ media, such as a *live* CD, may allow an antivirus program to find a Trojan and delete it. Updated anti-spyware programs are also efficient against this _____19_____. Most Trojans also _____20_____ in registries, and processes.

Words

Iliad['iliəd] *n.*《伊利亚特》(古希腊描写特洛伊战争的英雄史诗，相传为荷马所作)	**conceal**[kən'si:l] *v.*隐藏
siege[si:dʒ] *n.*围城，围攻	**payload**['pei,ləud] *n.*有效载荷
nightfall['naitfɔ:l] *n.*黄昏，傍晚	**dropper**['drɔpə] *n.*落下的人或物
	live[liv] *adj.*最新的

Phrases

start off	出发，动身

Software Engineering

Unit
11

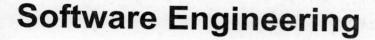

Part 1   Reading and Translating

Section A   Software Engineering

*Software engineering is the application of a systematic, **disciplined**, quantifiable approach to the development, operation, and maintenance of software. It encompasses techniques and procedures, often regulated by a software development process, with the purpose of improving the reliability and maintainability of software systems.* The effort is ***necessitated*** by the potential complexity of those systems, which may contain millions of lines of code.

The term software engineering was coined by Brian Randell and popularized by F.L. Bauer during the **NATO** Software Engineering Conference in 1968. *The discipline of software engineering includes knowledge, tools, and methods for software requirements, software design, software construction, software testing, and software maintenance tasks.* Software engineering is related to the disciplines of computer science, computer engineering, management, mathematics, project management, quality management, software ***ergonomics***, and systems engineering.

History

Software engineering has a long evolving history. Both the tools that are used and the applications that are written have evolved over time. It seems likely that software engineering will continue evolving for many decades to come.

1. 60 Year Time Line

- 1940s: First computer users wrote machine code by hand.
- 1950s: Early tools, such as *macro assemblers* and interpreters were created and widely used to improve productivity and quality.
- 1960s: Second generation tools like optimizing compilers and inspections were being used to improve productivity and quality. The concept of software engineering was widely discussed. First large scale (1000 programmer) projects. Commercial mainframes and custom software for big business. The influential 1968 NATO Conference on Software Engineering was held.
- 1970s: Collaborative software tools, such as Unix and code repositories, minicomputers and small business software appeared.
- 1980s: Personal computers and personal workstations became common. *Commensurate* rise of consumer software. Smalltalk: the first commercial OOP language/platform that is UI based, memory managed, **VM** Image based, scripted/incremental.
- 1990s: Object-oriented programming and agile processes like **extreme programming** gained mainstream acceptance. Computer memory capacity *sky-rocketed* and prices dropped *drastically*. These new technologies allowed software to grow more complex. The WWW and hand-held computers made software even more widely available.
- 2000s: Managed code and interpreted platforms such as Java, .NET, Ruby, Python and PHP made writing software easier than ever before. *Offshore* outsourcing changed the nature and focus of software engineering careers.

2. Current Trends in Software Engineering

Software engineering is a young discipline, and is still developing. The directions in which software engineering is developing include:

*Aspects help software engineers deal with quality attributes by providing tools to add or remove **boilerplate** code from many areas in the source code.* Aspects describe how all objects or functions should behave in particular *circumstances*. For example, aspects can add debugging, logging, or locking control into all objects of particular types. Researchers are currently working to understand how to use aspects to design general-purpose code. Related concepts include *generative* programming and templates.

Agile software development guides software development projects that evolve rapidly with changing expectations and competitive markets. Proponents of this method believe that heavy, document-driven processes (like TickIT, **CMM** and ISO 9000) are *fading* in importance. Some people believe that companies and agencies export many of the jobs that can be guided by heavy-weight processes. Related concepts include Extreme Programming and *Lean* Software Development.

Experimental software engineering is a branch of software engineering interested in devising experiments on software, in collecting data from the experiments, and in devising *laws* and theories from this data. Proponents of this method *advocate* that the nature of software is such that we can *advance* the knowledge on software through experiments only.

Model Driven Design develops *textual* and graphical models as primary design *artifacts*. Development *tools are available that use model transformation and code generation to generate well-organized code fragments that serve as a basis for producing complete applications.*

Software Product Lines is a systematic way to produce families of software systems, instead of creating a succession of completely individual products. This method emphasizes extensive, systematic, formal code reuse, to try to industrialize the software development process.

The Future of Software Engineering conference (**FOSE**), held at **ICSE 2000**, *documented* the *state of the art* of **SE** in 2000 and listed many problems to be solved over the next decade. The FOSE tracks at the ICSE 2000 and the ICSE 2007 conferences also help identify the state of the art in software engineering.

 Words

discipline[ˈdisiplin] *v.*通过教学和实践训练	**circumstance**[ˈsəːkəmstəns] *n.*环境，情况
necessitate[niˈsesiteit] *v.*使成为必需，迫使	**generative**[ˈdʒenərətiv] *adj.*有生产力的，生成的
ergonomics[ˌəːgəuˈnɔmiks] *n.*人类工程学，生物工程学	**fade**[feid] *v.*逐渐消失，衰弱
macro[ˈmækrəu] *n.*宏指令	**lean**[liːn] *adj.*简洁的，直接的
assembler[əˈsemblə] *n.*汇编程序	**law**[lɔː] *n.*规则，法则，规律
commensurate[kəˈmenʃərit] *adj.*成比例的，相称的	**advocate**[ˈædvəkit] *v.*提倡，主张
drastically[ˈdræstikəli] *adv.*激烈地，彻底地	**advance**[ədˈvɑːns] *v.*促进，建议
offshore[ˈɔ(ː)ʃɔː] *adj.*离岸的，国外的	**textual**[ˈtekstjuəl] *adj.*原文的，正文的，逐字的
aspect[ˈæspekt] *n.*方面	**artifact**[ˈɑːtifækt] *n.*人工制品
boilerplate[ˈbɔiləpleit] *n.*样板文件	**document**[ˈdɔkjumənt] *v.*证明，评述

Phrases

| with the purpose of | 以…为目的 | state of the art | 最新水平 |
| sky-rocket | 像火箭一样冲天 | | |

 Abbreviations

NATO	North Atlantic Treaty Organization	北大西洋公约组织
VM	Virtual Machine	虚拟机
XP	Extreme Programming	极限编程
CMM	Capability Maturity Model	能力成熟度模型
FOSE	Future Of Software Engineering Conference	软件工程前景讨论会
ICSE	International Conference on Software Engineering	软件工程国际会议
SE	Software Engineering	软件工程

Complex Sentences

1. **Original:** Aspects help software engineers deal with quality attributes by providing tools to add or remove boilerplate code from many areas in the source code.

   Translation: 通过提供工具以添加或移除来自源代码中多处的样板文件代码，"方面"能够帮助软件工程师处理质量属性。

2. **Original:** Development tools are available that use model transformation and code generation to generate well-organized code fragments that serve as a basis for producing complete applications.

   Translation: 使用模型转换和代码生成来产生组织良好的代码段的开发工具可用了，这些代码段可作为编制完整应用程序的基础。

Exercises

I. Read the following statements carefully, and decide whether they are true (T) or false (F) according to the text.

_____1. The potential complexity of software systems necessitates software engineering.

_____2. Custom software rose first from small business.

_____3. The development of hardware allows software to grow more complex and more widely available.

_____4. Systematic and formal code reuse is necessitated to industrialize the software development process.

_____5. Software engineering is subject to many other disciplines, such as computer science, project management, and systems engineering.

II. Choose the best answer to each of the following questions.

1. Which of the following emerged first in the history of software engineering?

   (A) Code repositories   (B) Commercial OOP language

   (C) Extreme programming   (D) Optimizing inspections

2. Which of the following is WRONG about the terminology used in software engineering?

   (A) Offshore outsourcing – Platform (B) Lean software development - Process

   (C) Macro assemblers – Tool   (D) Smalltalk - Language

3. Which of the following is RIGHT about the current trends in software engineering?

   (A) Aspects describe the behavior of all objects or functions in particular circumstances and design particular- purpose code for them.

   (B) Agile software development is a light and change -driven process.

   (C) Software Product Lines produce successions of individual software system products.

   (D) Experimental software engineering advances the knowledge on experiments through software.

III. Translating.

1. Software engineering is the application of a systematic, disciplined, quantifiable approach to the development, operation, and maintenance of software. It encompasses techniques and procedures, often regulated by a software development process, with the purpose of improving the reliability and maintainability of software systems.

2. The discipline of software engineering includes knowledge, tools, and methods for software requirements, software design, software construction, software testing, and software maintenance tasks.

⟫⟫ Section B   Software Development Process

A software development process is a structure imposed on the development of a software product. *Synonyms* include software life cycle and software process. There are several models for such processes, each describing approaches to a variety of tasks or activities that take place during the process.

➔ Process Activities/Steps

Software engineering processes are composed of many activities, notably the following:

1. Requirements Analysis

The most important task in creating a software product is extracting the requirements. Customers typically know what they want, but not what software should do, while incomplete, ambiguous or contradictory requirements are recognized by skilled and experienced software engineers. The system's services, constraints and goals are established by consultation with system users and then defined in a manner that is understandable by both users and development staff. This phase can be divided into: *feasibility study* (often carried out separately), requirements analysis, requirements definition and requirements **specification**. Frequently demonstrating live code may help reduce the risk that the requirements are incorrect.

2. Specification

Specification is the task of precisely describing the software to be written, possibly in a rigorous way. In practice, most successful specifications are written to understand and *fine-tune* applications that were already well-developed, although safety-critical software systems are often carefully specified prior to application development. Specifications are most important for external interfaces that must remain stable. A good way to determine whether the specifications are sufficiently precise is to have a third party review the documents making sure that the requirements and *use cases* are logically **sound**.

3. Design

System design   *Partitions* the requirements to hardware or software systems and establishes an overall system architecture. The architecture of a software system refers to an abstract representation of that system. Architecture is concerned with making sure the software system will meet the requirements of the product, as well as ensuring that future requirements can be addressed. The architecture step also addresses interfaces between the software system and other software products, as well as the underlying hardware or the host operating system.

Software design   Represents the software system functions in a form that can be transformed into one or more executable programs.

An important (and often overlooked) task is documenting the internal design of software for the purpose of future maintenance and enhancement. Documentation is most important for external interfaces and should be done throughout the whole development process.

4. Implementation and Testing

Implementation is the part of the process where software engineers actually write the code for the project.

Software testing is an integral and important part of the software development process. This part of the process ensures that bugs are recognized as early as possible. Software testing can be done on the following levels: unit testing, integration testing, system testing and validation testing.

There are many approaches to software testing. Reviews, walkthroughs or inspections are considered as static testing, whereas actually executing programmed code with a given set of test cases is referred to as dynamic testing. Testing tactics are traditionally divided into black box testing and white box testing. Black box testing treats the software as a black-box without any knowledge of internal implementation, while white box testing refers to the situation when the tester has access to the internal data structures and algorithms. There are also special methods existing to test non-functional aspects of software, such as performance testing, usability testing, and security testing.

5. Deployment and Maintenance

Deployment starts after the code is appropriately tested, is approved for release and sold *or otherwise* distributed into a production environment.

Software training and support is important because a large percentage of software projects fail because the developers fail to realize that it doesn't matter how much time and planning a development team puts into creating software if nobody in an organization ends up using it. People are often resistant to change and avoid venturing into an unfamiliar area, so as a part of the deployment phase, it is very important to have training classes for new clients of your software.

Maintaining and enhancing software to cope with newly discovered problems or new requirements can take far more time than the initial development of the software. *Not only may it be necessary to add code that does not fit the original design but just determining how software works at some point after it is completed may require significant effort by a software engineer.* About 60% of all software engineering work is maintenance, but this statistic can be misleading. A small part of that is fixing bugs. Most maintenance is extending systems to do new things, which in many ways can be considered new work.

➲Process Models

A decades-long goal has been to find repeatable, predictable processes or methodologies that improve productivity and quality. Some try to systematize or formalize the seemingly *unruly* task of writing software. Others apply project management techniques to develop software. Without project management, software projects can easily be delivered late or over budget. With large numbers of software projects not meeting their expectations in terms of functionality, cost, or delivery schedule, effective project management is proving difficult.

1. Waterfall Processes

The best-known and oldest process is the waterfall model, where developers follow these steps in order. They state requirements, analyze them, design a solution approach, *architect* a software framework for that solution, develop code, test, deploy, and maintain. After each step is finished, the process proceeds to the next step, just as builders don't revise the foundation of a house after the framing has been erected.

This is the central idea behind the waterfall model – time spent early on making sure that requirements and design are absolutely correct will save you much time and effort later. The waterfall model provides a structured approach; the model itself progresses linearly through discrete, easily understandable and explainable phases and thus is easy to understand; it also provides easily marked milestones in the development process. As well as the above, the waterfall model places emphasis on documentation (such as requirements documents and design documents) as well as source code, so that new team members or even entirely new teams should be able to *familiarize* themselves by reading the documents.

It is argued that the waterfall model in general can *be suited to* software projects which are stable (especially those projects with unchanging requirements, such as with *shrink wrap* software) and where it is possible and likely that designers will be able to fully predict problem areas of the system and produce a correct design before implementation is started. The waterfall model also requires that implementers follow the well made, complete design accurately, ensuring that the integration of the system proceeds smoothly.

2. Iterative Processes

*Iterative development **prescribes** the construction of initially small but ever larger portions of a software project to help all those involved to uncover important issues early before problems or faulty assumptions can lead to disaster.* Iterative processes are preferred by commercial developers because it allows a potential of reaching the design goals of a customer who does not know how to define what he wants.

Agile software development processes are built on the foundation of iterative development. To that foundation they add a lighter, more people-centric viewpoint than traditional approaches. Agile processes use feedback, rather than planning, as their primary control mechanism. The feedback is driven by regular tests and releases of the evolving software.

Agile processes seem to be more efficient than older methodologies, using less programmer time to produce more functional, higher quality software, but have the drawback from a business perspective that they do not provide long-term planning capability. In essence, they say that they will provide the most bang for the buck, but won't say exactly when that bang will be.

While iterative development approaches have their advantages, software architects are still *faced with* the challenge of creating a reliable foundation upon which to develop. Such a foundation often requires a fair amount of *upfront* analysis and prototyping to build a development model. The development model often relies upon specific design patterns and entity relationship diagrams (**ERD**). Without this upfront foundation, iterative development can create long term challenges that are significant in terms of cost and quality.

In addition, more and more software development organizations have implemented process methodologies. The Capability Maturity Model Integration (**CMMI**) is one of the leading models and based on best practice. Independent assessments grade organizations on how well they follow their defined processes, not on the quality of those processes or the software produced. CMMI has replaced CMM. There are dozens of others, with other popular ones being ISO 9000, ISO 15504, and Six Sigma.

Formal methods

Formal methods are mathematical approaches to solving software (and hardware) problems at the requirements, specification and design levels. Examples of formal methods include the B-Method, Petri nets, **RAISE** and **VDM**. Various formal specification notations are available, such as the Z notation. More generally, *automata* theory can be used to build up and validate application behavior by designing a system of finite state machines (**FSM**). Methodologies based on FSM allow executable software specification and by-passing of conventional coding. Formal methods are most likely to be applied in *avionics* software, particularly where the software is safety critical.

Another emerging trend in software development is to write a specification in some form of logic (usually a variation of **FOL**), and then to directly execute the logic as though it were a program. The **OWL** language, based on Description Logic, is an example. There is also work on mapping some version of English (or another natural language) automatically to and from logic, and executing the logic directly. Examples are Attempto Controlled English, and Internet Business Logic, which does not seek to control the vocabulary or syntax. A feature of systems that support *bidirectional* English-logic mapping and direct execution of the logic is that they can be made to explain their results, in English, at the business or scientific level.

 Words

synonym['sinənim] *n.*同义词	**familiarize**[fə'miljəraiz] *v.*使熟悉，使被认知
specification[.spesifi'kei∫ən] *n.*规格，规约，说明书	**prescribe**[pris'kraib] *v.*指定，规定
sound[saund] *adj.*正确的，可靠的，合理的	**upfront**[ʌpfrʌnt] *adj.*在前面的，在最前面的

partition[pɑ:'tiʃ ən] *v.*划分，把…分成部分	**automata**['ɔːtəmətə] *n.*自动机
unruly[ʌn'ru:li] *adj.*难以控制的	**avionics**[,eivi'ɔniks] *n.*航空电子技术
architect['ɑ:kitekt] *v.*设计，建造	**bidirectional**[,baidi'rekʃ ənəl] *adj.*双向的

 Phrases

feasibility study	可行性研究	be suited to	适合于
fine-tune	调整，微调	shrink wrap	作为套装软件出售
use case	用例	be faced with	面临
or otherwise	或者相反	formal method	形式化方法
end up	告终，结束		

 Abbreviations

ERD	Entity Relationship Diagrams	实体关系图
CMMI	Capability Maturity Model Integration	能力成熟度模型集成
RAISE	Rigorous Approach to Industrial Software Engineering	工业软件工程的严格方法
VDM	Vienna Development Method	维也纳开发方法
FSM	Finite State Machines	有限状态机
FOL	First-Order Logic	一阶逻辑
OWL	Web Ontology Language	网络本体语言

Complex Sentences

1. **Original:** Software training and support is important because a large percentage of software projects fail because the developers fail to realize that it doesn't matter how much time and planning a development team puts into creating software if nobody in an organization ends up using it.

   Translation: 软件培训和支持很重要，这是因为大多数软件项目的失败是由于开发者没有认识到：如果机构中没有人最终使用它，那么开发团队无论投入多少时间和计划到开发软件中都无济于事。

2. **Original:** Not only may it be necessary to add code that does not fit the original design but just determining how software works at some point after it is completed may require significant effort by a software engineer.

   Translation: 不仅是可能需要添加与最初设计不相符的代码，且仅仅确定软件在其完成后的某一点上如何工作都可能需要软件工程师的大量努力。

Exercises

I. Read the following statements carefully, and decide whether they are true (T) or false (F) according to the text.

____1. Process models describe approaches to a variety of tasks or activities that take place during the software development process.

____2. The iterative model has replaced the traditional waterfall model due to its higher efficiency.

____3. CMMI can guide software development organizations to follow their defined processes.

____4. Formal methods are mathematical approaches to solving software problems which are most likely to be applied in mathematics area.

____5. Reducing even bypassing the workload spent on coding may be an obvious part of the software engineering development in future.

II. Choose the best answer to each of the following questions.

1. Which of the following is RIGHT about the activities composing software engineering processes?

   (A) The initial requirements from customers should be complete, unambiguous and consistent.

   (B) Documentation is important for external interfaces and should begin in the phase of design.

   (C) An overall system architecture should be established before the detailed design.

   (D) Testing ensures fixing bugs as early as possible and completely establishing the correct software.

2. Which of the following is WRONG about the differences between the waterfall model and the iterative model?

   (A) The waterfall model places more emphasis on documentation than the iterative model.

   (B) The iterative model is better for software projects where requirements are inexplicit initially.

   (C) The waterfall model is more suited to software projects which are stable and predictable.

   (D) The iterative model is more efficient and successful due to no needing any upfront planning.

3. Why does the author state that "*Frequently demonstrating live code may help reduce the risk that the requirements are incorrect*"?

   (A) Live code can define the requirements more understandably, replacing the requirements specification.

   (B) Live code can help user to recognize and validate the system's requirements besides oral consultation.

   (C) Live code can demonstrate and design software as early as possible bypassing the requirements analysis.

(D) Live code can demonstrate and implement software as early as possible bypassing the requirements analysis.

III. Translating.

1. This is the central idea behind the waterfall model - time spent early on making sure that requirements and design are absolutely correct will save you much time and effort later.

2. Iterative development prescribes the construction of initially small but ever larger portions of a software project to help all those involved to uncover important issues early before problems or faulty assumptions can lead to disaster.

Part 2 **Simulated Writing: Resume**

⊖ Introduction

A resume - sometimes called a "Curriculum Vitae" or "CV" - should be a concise and factual presentation of your credentials and is a summary of your education and work experience. A good resume demonstrates how your skills and abilities match up with the requirements of a job.

As an invaluable marketing tool, resumes give you the opportunity to introduce yourself to a potential employer. A resume is not an exhaustive list - if it's too long it probably won't be readable at all. The best resumes are usually no longer than one or two pages.

You should change your resume with every job application so that it lists the skills and experience you have that are most relevant to the job you're applying for.

⊖ Things to Include on a Resume

While evidences like your academic record and work experience are typical ways to show how suited you are to a particular job, these are not the only things you should include. There are other ways to demonstrate that you are the right person for the job, including the following points.

- Personal traits – you're an honest worker, you enjoy being part of a big team
- Strengths – things you're good at or enjoy
- Experiences – volunteering or extracurricular activities
- Key responsibilities and achievements – any awards or recognition that you've received

⊖ Resume Structure

The structure of your resume will vary depending on your work experience and education and training background.

A resume usually sets out information in a reverse order. Your most recent work experience and study details should be first on the list. A typical resume includes:

1. Identifying Information

Full name and contact details including address, telephone number(s) and email address. It is important to provide all of this information to a prospective employer. It is always placed at the top of the resume. Be sure to make your name a little larger so it stands out to the employer.

2. Career Objective (Optional)

A short but succinct description of the type of position(s) telling the employer what type of job you are currently seeking and showing that you've given thought to your future career. It is a guiding statement that helps the employer to direct the resume to the appropriate person for the appropriate job while giving your resume a focus.

3. Education and Training

A summary of your education and training history, starting with your most recent studies, making sure you include all training that's relevant to the job you're applying for.

4. Employment History

Start with your most recent work history and work backwards chronologically, listing the name of the employer, your job title, the dates you worked there, and your responsibilities, tasks and achievements. Make sure you include everything that's relevant to the job.

5. Skills and Abilities

A list of skills you're good at. These can be general skills or skills specific to a particular job. List them under broad headings such as "Communication" and "Teamwork".

6. Interests (Optional)

A list of your hobbies and interests; this gives employers more information about you and also shows other areas of your life where you've gained experience such as teamwork and commitment.

7. Referees

List people who can talk about how good a worker you are. Make sure you get their permission before including them on your resume. List their name, company name occupation, and contact details.

⊖Other Useful Information

- Limit to one or two pages. You do not have to state everything you have done on a resume, but everything you state must be true.
- Determine the type of layout that works best for your experience. You may use bolding to emphasize key skills and accomplishments.
- Tailor your resume to the type of position to which you are applying. Decide what you want your resume to convey about your abilities.
- Maintain a consistent writing style.
  a) Do not use "I" or "my."
  b) You may use complete or fragmented sentences as long as the meaning is clear and style is consistent.
  c) Start each description with an action word. Use present tense verbs when referring to current activities. Use past tense verbs when referring to past activities.

- Emphasize outcome, accomplishments and breadth of responsibility.
- Be concise and clear in your descriptions. Do not try to impress employers with the use of complicated or confusing words.
- Make sure there are no typing, spelling or grammar errors and have someone proofread your resume.
- Do not use contractions and make sure you define abbreviations or acronyms.
- Make sure the resume is well laid out, easy to read:
  a) Choose a font that is easy to read (examples are Arial, Tahoma, Times New Roman or Verdana), and a standard size (10 or 12 point).
  b) Be consistent. For example, if your headings are in bold type, all headings should be in bold. Each entry should follow a uniform format.

- Print final copies of your resume on quality paper and make sure the paper photocopies well.
- In most cases a cover letter should be attached to your resume.

⊖Sample

Mark Li
P.O. Box 112
37 Xueyuan Rd. Haidian District
Beijing, PR China
(010) 1234-5678
mark@bh.edu.cn

OBJECTIVE	An entry-level position in the field of software engineering
EDUCATION	Beihang University, Beijing
	Bachelor of Engineering, May 2008
	Major: Software Engineering
	GPA: 3.84/4.00
CAREER RELATED EXPERIENCE	**System Designer & Programmer**, 2008
	Digital Earth & GIS Lab, Beihang University, Beijing
	• Designed and developed UI of a platform of Digital Earth sponsored by National Fund
	• Designed and developed a 3D simulation system for airplane
	System Designer& Programmer, 2007 - 2008
	College of Software, Beihang University, Beijing
	• Designed and developed Resource Management System for College Financial Office
	• Designed and developed Information Management System for College Library
ADDITIONAL EXPERIENCE	**Intern Market Research Analyst,** 2008
	FE Software, Inc., Beijing
	• Did research and analysis on global and domestic markets of outsourcing software in a complete English environment
	• Wrote a business report and plan for the company in English
	Teaching Assistant, 2006 - 2007
	College of Software, Beihang University, Beijing
	• Assisted teaching and activity organizing, communicated with students on study and career planning
	• Assisted organizing a mock interview cooperating with MOTOROLA in 2007
SKILLS	**Language:** CET-6, fluency in written and spoken English
	Computer: Having rich and systematic knowledge in computer and software engineering domain
	Be familiar with Oracle, C#, and Java programming
	Mastering a variety of UI design software
HONORS/ACTIVITIES	First Award of "Student Scholarship", Beihang University, 2007
	Third Award of "Creative Technology", Beihang University, 2007
REFERENCES	Available upon request

Part 3 **Listening and Speaking**

⫸ **Dialogue: Using Object-Oriented Analysis and Design Method**

(Henry, Mark, and Sophie are making a project planning for a library information management system as their course project.)

Mark	Having gathered the requirements of the system, we can progress further!
Sophie	Yes. The next step is to produce analysis modeling and design implementation specifications.
Henry	Based on our *acquaintance with* this system through the requirements gathering, I think we can use object-oriented analysis and design method for easier and more obvious mapping from real world entities to system objects.
Mark	It's a good idea. Using object-oriented analysis and design method, we can **model** the system as a group of interacting objects and more sense for object-oriented languages.[1]
Sophie	*With regard to* object-modeling techniques, there are a number of different notations for representing various models showing the static structure, dynamic behavior, and run-time deployment of these collaborating objects.[2] Which methodologies or tools shall we use?
Mark	How about UML, *by right of* its being standardized and general- purpose? [3]
Sophie	I completely agree with you, Mark.
Henry	Me too. Then we can focus on the tasks in each stage.
Mark	Yes. In the phase of object-oriented analysis, following the written requirements statement and applying object-modeling techniques, we look at the problem domain and analyze the functional requirements for a system with the aim of producing a conceptual model.
Sophie	What kinds of artifact should be created embodying the conceptual model?
Henry	That will typically be presented as a set of use cases, one or more UML class diagrams, and a number of interaction diagrams. It may also include some kind of user interface *mock-up*.

Mark	Next, the output of analysis provides the input for design. Object-oriented design elaborates the analysis models to produce implementation specifications. The concepts in the analysis model are mapped onto implementation classes and interfaces. The result of object-oriented design is a model of the solution domain, a detailed description of how the system is to be built.
Henry	The deliverables of object-oriented design will be in general a set of sequence diagrams and class diagram.[4]
Sophie	So that, do you mean we must perform the analysis completely before beginning the design?
Mark	Not always. In practice, one activity can feed the other in a short feedback cycle through an iterative process, so analysis and design may occur in parallel. Both analysis and design can be performed little by little, and the artifacts can be continuously grown instead of completely developed *in one shot.*
Henry	In addition, we may use some ***acknowledged*** design concepts such as design patterns and application frameworks, as well as some design principles, for example, dependency injection and composite reuse principles to refine our design.[5]

Exercises

Work in pairs, and make up a similar conversation by replacing the scenarios with other material in the below.

Using Structured Analysis and Design Method

[1] Structured analysis and design method features a top-down, hierarchical approach that tends to generate well-organized systems. Its step-by-step approach can simplify early detection of design flaws, improve the ability to modify programs, make clear and complete documentation easily, and *break up* problem into modules to improve testing and to allow development by multiple design teams more easily.

[2] There are many structured analysis and design methodologies that can be used to implement programs.

[3] One of the most general techniques is a top down, structured flowchart methodology.

[4] "Structured analysis and design" is divided into two components: structured analysis and

structured design. More generally, structured analysis transforms the abstract problem into a feasible logical design, while structured design concentrates on converting the logical design into a physical information system.

(1) Structured Analysis

A top-down analysis begins with a carefully structured textual description of the problem to be solved. From the text description, a top-level flowchart is created. This describes the sequence of processes to be carried at a high level with few details. Each process block in this high level flowchart is then broken down into further detailed blocks through another flowchart that describes how the higher-level process is to be implemented. In a complex design, many levels may be required before reaching a sufficiently detailed level of description to write the code.

Primary output:

- Data Flow Diagram (with data dictionary and mini-specification): concerned about data flow and functional *decomposition* to describe how the system works.
- Entity Relationship Diagram: The data are separated into entities and relationships which are the associations between the entities.

(2) Structured Design

Structured design *is concerned with* physical design based on the results of structured analysis and views the world as a collection of modules with functions that share data with other modules.

Primary output:

- Structure Chart

[5] Structured design is also based on design *heuristics* like object-oriented design, such as coupling, cohesion, encapsulation, modularity, etc.

 Words

model['mɔdl] *v.*建模	**decomposition**[,di:kɔmpə'ziʃən] *n.*分解
acknowledged[ək'nɔlidʒid] *adj.*公认的，被普遍认可的	**heuristics**[hjuə'ristiks] *n.*启发法

 Phrases

be acquainted with	了解，结识	**in one shot**	立刻，马上
with regard to	关于，对于	**break up**	分解
by right of	由于，因为	**be concerned with**	参与
mock-up	模型，原型		

Listening Comprehension: Extreme Programming

Listen to the passage and the following 3 questions based on it. After you hear a question, there will be a break of 10 seconds. During the break, you will decide which one is the best answer among the four choices marked (A), (B), (C) and (D).

Questions

1. When did the XP's practice and methodology begin to be used?
2. What is the main objective of XP?
3. Which of the following does NOT accord with the XP's essential value?

Choices

4. (A) 1960s   (B) 1980s   (C) 1990s   (D) 2000s
5. (A) Reducing the cost of design change   (B) Reducing the cost of requirement change
   (C) Reducing the cost of testing change   (D) Reducing the cost of delivery change
6. (A) Emphasizing customer involvement   (B) Promoting team work
   (C) Keeping the design integrated   (D) Getting feedback by software testing daily

 Words

---

inescapable[,inis'keipəbl] *adj.*不可避免的，不可忽视的
practitioner[præk'tiʃənə] *adj.*从业者，实践者
DaimlerChrysler['daimlə'kraislə] *n.*戴姆勒–克莱斯勒
refactoring['fæktəriŋ] *n.*重构

Dictation: Unified Modeling Language (UML)

This passage will be played THREE times. Listen carefully, and fill in the blanks with the words you have heard.

Unified Modeling Language (UML) is a standardized ____1____ modeling language in the field of software engineering. It is a ____2____ language for visualizing, specifying and constructing the ____3____ of a software-intensive system. UML includes a set of graphical ____4____ techniques and offers a standard way to write a system's ____5____ and abstract models, including conceptual things such as business processes and system functions as well as concrete things such as programming language statements, database schemas, and ____6____ software components. UML can be used with all processes, throughout the software development life ____7____ , and across different ____8____ technologies.

UML 2.0 has 13 types of diagrams divided into three ____9____:Six diagram types represent structure application structure, three represent general types of ____10____ , and four represent different aspects of _____11_____ , representing three different _____12_____ of a system model: functional requirements *view*, static structural view and dynamic behavior view.

Under the technical _____13_____ of the three methodologists, James Rumbaugh, Grady Booch and Ivar Jacobson, who were collectively referred to as the Three *Amigos*, UML 1.0 specification draft was _____14_____ to the Object Management Group (**OMG**) in January 1997. Today, UML has _____15_____ significantly since the first version of UML, _____16_____ by the UML 2.0 major revision that was _____17_____ by the OMG in 2005.

UML is not a development method by itself; however, it was designed to be _____18_____ with the leading object-oriented software development methods of its time (for example Rumbaugh's Object Modeling Technique (**OMT**), Booch method). Since UML has _____19_____ , some of these methods have been *recast* to take _____20_____ of the new notations, and new methods have been created based on UML. The best known is IBM Rational Unified Process (**RUP**). There are many other UML-based methods like Abstraction Method, Dynamic Systems Development Method, and others, designed to provide more specific solutions, or achieve different objectives.

Words

view[vju:] *n.*视图	**recast**['ri:'kɑːst] *v.*重铸，改写
amigo[ə'miːgəu] *n.*朋友	

Abbreviations

OMG	Object Management Group	对象管理组织
OMT	Object Modeling Technique	对象建模技术
RUP	Rational Unified Process	Rational 统一过程

The Future of Computer

Part 1   Reading and Translating

Section A   The Future of Computing (1)

Life goes in circles, or in spirals to be more precise. Thus to get a glimpse of the future you perhaps should take a look at the past. Computing has become an inseparable component of our lives and therefore is a subject to the same law of cycles and spirals. So let's start in the 1940s, at the dawn of electronic computing when ENIAC was a *pinnacle* of scientific engineering.

Initially occupying whole buildings, then scaling down to individual rooms and towering boxes computers of 1950s,'60s, and '70s were essentially mainframes: Large and powerful, special-purpose, accessible to few. This centralized approach to computing changed dramatically in the late '70s and early '80s with the introduction of microcomputers such as Apple II and IBM PC. *All of a sudden* computers ceased to be shared resources built for a particular purpose, and became personal tools for facilitating general-purpose tasks instead. Moreover, they began to occupy our desktops, bedrooms and closets throughout '80s and '90s.

This trend of decentralized computing met a subtle reverse in '90s when the Internet and World Wide Web provided a means for integrating decentralized computational resources into a unified client-server environment. In reality, what seems like 60 years of technological advancement represents a full evolutionary cycle: We started with shared computational resources occupying rooms of equipment and through brief desktop detour arrived at shared computational resource model built on the backbone of Intranet/Internet. There is unquestionable numerical difference between what we have 30 years ago and what we have now in the sense that computers now are being used by much larger population and for a far wider range of tasks. There is a characteristic difference too: We kept our desktop PCs and we use them for more than mere terminals. In fact computing did not just come a full circle; it came a loop of spiral: We have not really come back to good old centralized computing but rather to arrive at distributed computing model. Although *a bulk of* work may be done by centralized resources such as servers providing computational services, our desktop PCs and client workstations independently handle a ***multitude*** of tasks.

Our equipment rooms are different too from what we had decades ago: Instead of one large computer modern day data centers are filled with hundreds and thousands of servers both *rack mount* and ***blades***. Such evolutionary change parallels that of the evolution of electronic components: Discrete elements were gradually replaced with integrated circuits just as individual mainframes are now replaced with blade server racks. *Extrapolating the parallel further we may expect even tighter-integrated "microblades" to arrive in the near future when we master computer integration to the same degree as we have mastered integrated circuits.*

Evolutionary Changes

Internally, computers are undergoing cyclical evolutionary changes as well: CPUs gradually evolved from spending tens if not hundreds of cycles on individual instruction to just one cycle per instruction (*scalar* architecture, for example). Then advent of additional execution units allowed CPUs processing several instructions per clock cycle thus exploiting instruction level parallelism (superscalar architecture, for instance). Later several CPUs were ***crammed*** on motherboard (multi-processor architecture). Now several CPUs are ***fused*** together in a single multi-core package, and several such multi-core chips can be installed on a single motherboard. So in the end a data center is filled with clusters of stacks of server blades with each blade ***sporting*** one or more multi-core super-scalar CPUs. So we have at least five different levels of integration: Cluster, Server, CPU, Core, and Execution unit, with four levels typically found on desktop PCs. But why do we need this complexity and how did it come into being?

As computer clock speed increased from kilohertz to gigahertz so did our imagination and understanding of what can be done with this computational power to serve our needs; for example, provide entertainment (at home) and boost productivity (which is a practical reason for computers in the workplace). Originally computers were designed to serve a clearly defined special purpose and therefore were meant to perform specific single task. When computer power grew beyond immediate needs multi-tasking was invented to allow multiple users access to the spare computational resources. But when the hardware costs were reduced to consumer level personal computers came out, and they quite naturally were designed to be single-tasking.

The first mass-produced personal computers were quite slow and the need for performance increase of microcomputer processors was justified at first: We wanted good response from desktop applications and occasionally we wanted to play *arcade* games. And 4.77 MHz of PC XT was not always good enough for the purpose. So as CPU power grew to meet specific tasks we wanted our PCs to perform it became too much for general tasks such as text editing or spread-sheeting. That extra power led to the adoption of multi-tasking operating systems on desktop and personal computers. We had extra power and we wanted to do something with it.

Ironically, mass adoption of multi-tasking operating system on PCs (Microsoft Windows, for instance) *coincided* with the introduction of graphical user interface. Thus formerly more or less satisfactory CPU performance became vastly inadequate and spurred a race to increase CPU performance necessary to compensate for inefficient software. The whole transition from single-threading text-based DOS programs to graphic interfaces and multithreading Windows resulted in unprecedented *bloating* of software code and general system slow down due to lacking graphic and disk I/O performance. Erroneous programming paradigms such as dynamically loaded libraries, dynamic memory allocation, shared components, inefficient object-oriented programming, and multi-layered libraries also greatly contributed to the slow down. All these inefficiencies instantly justified the need for further CPU performance increases. Now we needed faster computers just to run our operating system and new versions of old software burdened with graphic user interfaces.

Thus *paradoxically* desktop computers of '80s initiated a major leap in Wirth's Law which states that software is getting slower faster than computers are getting faster. Perhaps the first loop of Wirth's Law spiral was objective: Initial **CGA** and **EGA** hardware and CPU performance of 12-16 MHz was barely enough for running programs with complicated graphic interfaces. However, further *unraveling* of Wirth's Law was completely subjective in the sense that slowdown of software that occurred further resulted from our attempts to boost programmer productivity by employing various "coding techniques" that promised simplicity at the cost of efficiency.

Software Performance versus Developer Productivity

Indeed the first mainframes were programmed directly in machine code and a bit later in assembler. Severe memory limitations and simplicity of original instruction set (PDP 11 is a classical example) and relative simplicity programming tasks *at hand* resulted in highly efficient if not bare bone code that, unfortunately, was difficult to write. When in late 1950s computers became fast enough to relieve some of the coding burden from the shoulders of programmers high level languages were developed such as Ada, Algol, Fortran and C. While sacrificing code efficiency big time these high-level languages allowed us to write code faster and thus extract more productivity gains from computers.

As time passed we kept sacrificing software performance in favor of developer productivity gains first by adopting object-oriented languages and more recently settling with garbage-collected memory, runtime interpreted languages and "managed" execution. It is these "developer productivity" gains that kept the pressure on hardware developers to *come up with* faster and faster performing processors. So one may say that part of the reason why we ended up with gigahertz-fast CPUs was "dumb" (lazy, uneducated, expensive – pick your favorite *epithet*) developers. Even now major OS release (such as Windows Vista) seems to be main reason for computer upgrades because new software runs slower doing the same tasks as the older software it replaces. Of course, the new software usually does much more than the old one. So an objective reason for faster computers is higher expectations and expanded feature set of the new software. After all there are some mission-critical applications that really demand performance. Database applications on server side and games on desktop size are good examples of such apps that drive hardware development towards faster performance objectively.

Still, if you look at your desktop OS now it is unbelievably bloated. For instance, when running Windows XP under normal circumstances you will easily count 50 processes, 500 threads and only about 5-10 percent CPU utilization. Thus a 10 times less powerful CPU would have equally well satisfied the requirements for browsing and typing. Yet computers in general and CPUs in particular keep getting faster and faster driven both by developer productivity needs and the requirements of mission-critical applications.

Incredibly a new factor *kicked in* that is threatening to *curb* raw clock speed increases – *runaway* power consumption. It is not unusual for a modern CPU to dissipate in excess of 100 Watts, which in the case of data centers translate into tens of millions of direct power and cooling costs. So on one hand we have a habit (but rarely a need) for higher performance and on the other hand we have a *looming fossil* fuel crisis, global warming and rising energy prices. Shall we finally stop *racing* the clock speed?

Apparently, the trend for higher clock speed has already been reversed when cooler yet efficiently running AMD's Athlon processor managed to win sizeable market share from hotter and faster by clock speed Intel's Pentium 4. So how are we going to *keep up with* performance demands without *liberal* increases in CPU clock frequency?

⊖ Maintaining Performance

Well, there are many ways to maintain performance. The first one – exploitation of instruction level parallelism – resulted in creation of super-scalar processors that we see today. Theoretically any modern CPU whether from Intel, AMD, IBM, or Sun can process and **retire** multiple instructions per cycle due to multiple parallel internal execution units. *Funny enough, instruction-level parallelism does not yet allow **sustained** performance of **substantially** more than 1 instruction per cycle (**IPC**) on general **benchmarks** due to memory latency and **branch** misprediction **penalty** that stalls even the fastest CPUs more than half the time.* Only highly-optimized tests or special-purpose code is capable of 3x to 5x performance boost warranted by multiple execution units. Practical gains due to architectural improvements of cache coherency or branch prediction *amount to* mere 5 percent in general. Long *multi-stage* execution **pipelines** that were developed to achieve higher clock speeds and inadequate memory performance created a situation when CPU can process data faster than the data can be supplied. So the trend for higher clock speed has already reversed in favor of shorter pipelines and better memory throughput. The best example of pipeline shortening is UltraSparc T1 processor with its six stage pipeline as opposed to 31-stage Pentium 4 models (Athlon XP has 10-stage pipeline and Intel's new "Woodcrest" server chip as only 14). Extrapolating the trend it is reasonable to expect CPU frequency to roughly remain the same while the CPU performance will increase due to pipeline shortening and emphasis on memory subsystem performance improvements.

Still, there is a hard limit for instruction-level parallelism, which makes it difficult in practice to keep individual execution units inside a CPU busy. Thus to improve CPU efficiency two alternative approaches are currently being **pursued**. One approach is super-threading (or hyper-threading if we use Intel's terms), which allows CPU to process several parallel threads simultaneously switching from one thread to another when a stall occurs. UltraSparc T1 takes this approach to extreme by executing four threads on each core (with 32 threads on 8-core chip), switching threads in *round-robin* manner when a stall occurs. While super-threading certainly boosts performance of multi-threaded applications **speculative** threading is pursued for improving performance of critical single-threaded applications. Intel is highly involved in speculative threading research and offers a Mitosis technology that designates threads most suitably for speculative execution with the help of compilers. AMD is developing similar technology, although the company is more *tight-lipped* about it. Still many rumors are circulating about AMD's **clandestine** "inverse hyper-threading" technology **allegedly** capable of uniting two individual CPU cores into a single CPU super-core CPU that would **crunch** single-threaded applications with a considerable performance boost. Yet the only piece of evidence on AMD's involvement with speculative threading that so far surfaced is **infamous** U.S. patent # 6,574,725, looking like hardware support for speculative threading in the vein of Intel's Mitosis. So with clock-speed increases effectively curbed by power consumption concerns most likely performance gains on upcoming CPUs would be due to super-threading (server chips) and speculative-threading (desktop chips).

There is another approach for boosting instruction-level parallelism, which has been pursued on and off by various commercial and government entities. This means very-large instruction word (**VLIW**) or explicitly-parallel instruction set (**EPIC**) computing. First successful application of VLIW concept can be tracked back to early 1980s when a group of Russian engineers lead by Boris Babayan (who is now an Intel fellow) developed a series of Elbrus supercomputers as a part of the anti-ballistic missile defense system deployed around Moscow. *Massive performance gains warranted by proper application of VLIW concept allowed Elbrus machines to overcome manufacturing and technological limitations and beautifully serve their purpose.* Remember that these were a special-purpose computers running hand-optimized code.

 Words

pinnacle[ˈpinəkl] *n.*高峰，尖端	**fossil**[ˈfɔsl] *adj.*过时的，陈旧的
detour[ˈdiːtuə] *n.*迂回，弯路	**race**[reis] *v.*与…赛跑，使全速行进
multitude[ˈmʌltitjuːd] *n.*大量	**liberal**[ˈlibərəl] *v.*自由的，大量的
blade[bleid] *n.*刀片	**retire**[riˈtaiə] *v.*停止使用，使报废
scalar[ˈskeilə] *n.*标量	**sustained**[səsˈteind] *adj.*持续不变的，相同的
cram[kræm] *v.*填满	**substantially**[səbˈstænʃ(ə)li] *adv.*基本上
fuse[fjuːz] *v.*合并，结合一起	**benchmark**[ˈbentʃˌmɑːk] *n.*基准
sport[spɔːt] *v.*浪费	**branch**[brɑːntʃ] *n.*分支程序
arcade[ɑːˈkeid] *n.*游乐中心，娱乐厅	**penalty**[ˈpenlti] *n.*损失
coincide[ˌkəuinˈsaid] *v.*同时发生	**pipeline**[ˈpaipˌlain] *n.*管道
bloat[bləut] *v.*膨胀	**pursue**[pəˈsjuː] *v.*进行，继续
paradoxically[ˌpærəˈdɔksikəli] *adv.*荒谬地，自相矛盾地	**speculative**[ˈspekjulətiv] *adj.*推测的
unravel[ʌnˈrævəl] *v.*阐明，解决	**clandestine**[klænˈdestin] *adj.*秘密的
epithet[ˈepiθet] *n.*绰号，称号，表述词语	**allegedly**[əˈledʒidli] *adj.*声称的
curb[kəːb] *v.*抑制	**crunch**[krʌntʃ] *v.*运行，处理
runaway[ˈrʌnəwei] *adj.*失控的	**infamous**[ˈinfəməs] *adj.*声名狼藉的
loom[luːm] *v.*迫在眉睫	

 Phrases

all of a sudden	突然	**keep up with**	赶上，不落后于
a bulk of	大部分	**amount to**	总计
rack mount	机架固定件	**multi-stage**	多级的

at hand	在手边，在近处	round-robin	一系列
come up with	提出	tight-lipped	紧闭嘴巴的，沉默的
kick in	踢掉砸开		

Abbreviations

CGA	Color Graphics Array	彩色图形阵列
EGA	Enhanced Graphics Array	增强型图形阵列
IPC	Instruction Per Cycle	每周期指令
VLIW	Very-Large Instruction Word	超长指令字
EPIC	Explicitly Parallel Instruction Computing	显式并行指令计算

Complex Sentences

1. **Original:** We started with shared computational resources occupying rooms of equipment and through brief desktop detour arrived at shared computational resource model built on the backbone of Intranet/Internet.

   Translation: 我们从占用多个设备房间的共享计算资源开始，经过短暂的桌面"弯路"，到达构建在企业内网/因特网的主干网上的共享计算资源模型。

2. **Original:** Funny enough, instruction-level parallelism does not yet allow sustained performance of substantially more than 1 instruction per cycle (IPC) on general benchmarks due to memory latency and branch misprediction penalty that stalls even the fastest CPUs more than half the time.

   Translation: 十分有趣的是，指令级并行尚不允许在通用基准上的每个周期多于一条指令的持续性能，这是由于内存等待时间以及甚至会使最快的CPU延迟多于一半时间的分支程序错误预测损失的原因。

Exercises

I. Read the following statements carefully, and decide whether they are true (T) or false (F) according to the text.

____1. When computer power growing boosted multi-tasking, the hardware cost reducing spurred personal computers quite naturally designed for single-tasking.

____2. In superscalar architecture, several CPUs are fused together in a single multi-core package, and several such multi-core chips can be installed on a single motherboard.

____3. Developer productivity gains appeal to higher performing processors.

____4. Runaway power consumption has put a curb on raw clock speed increases.

____5. Now, Practical gains have amounted to 3x to 5x Performance boost warranted by multiple execution units.

II. Choose the best answer to each of the following questions.

1. Which of the following does best argue the author's point that *computing is a subject to the same law of cycles and spirals*?
   (A) We started with mainframes which were accessible to few and now arrive at microcomputers used by much larger population.
   (B) We started with computers designed to serve a special purpose and now arrive at microcomputers used for general purpose.
   (C) We started with shared centralized computing and through decentralized desktop, now arrive at shared distributed computing model.
   (D) We started with discrete electronic components and now arrive at integrated circuits with much less room.

2. Which of the following is WRONG about the relationships of software performance, developer productivity and hardware performance?
   (A) Continuously expanded feature set of the new software requires higher hardware performance
   (B) Developer productivity needs to drive faster hardware performance
   (C) Higher hardware performance compensates for inefficient software
   (D) Programmer productivity boost results in higher software performance

3. Which of the following is the most direct reason for bringing in super-threading and speculative-threading approaches?
   (A) To solve the problem of instruction-level parallelism boost
   (B) To solve the problem of multiple execution units schedule
   (C) To solve the problem of pipeline shortening
   (D) To solve the problem of memory throughout improvement

III. Translating.

1. Extrapolating the parallel further we may expect even tighter-integrated "microblades" to arrive in the near future when we master computer integration to the same degree as we have mastered integrated circuits.

2. Massive performance gains warranted by proper application of VLIW concept allowed Elbrus machines to overcome manufacturing and technological limitations and beautifully serve their purpose.

▶▶ Section B   The Future of Computing (2)

⊙VLIW

Commercial applications of VLIW concept in the U.S. were less successful: Multiflow Computer went down in 1990 and Intel's EPIC/Itanium adventure of late 90s and today proved to be far from successful. The reason for VLIW failure on general-purpose computers is the lack of compilers, cross-compilers and automatic code optimization techniques. Intel is still heavily involved in *honing* EPIC compilers for Itanium (with Babayan's current team and Intel's Israeli's office heavily involved). *Yet the state of current technology is such that current VLIW/EPIC compilers are not yet good enough for general purposes and therefore theoretically possible performance gains are almost never achieved (VLIW processors can execute as many as 32 instructions in parallel if a compiler can find and schedule that many).* More recent attempt by Transmeta was also unsuccessful for the same reason, although it's new CPU looks more promising than *flopped* Crusoe. Still, with Itanium disappointment *tarnishing* commercial VLIW *prospects* perhaps permanently we are unlikely to see more general-purpose VLIW computers, instead to see them in *niche* markets employed for solving a very limited set of special-purpose tasks.

Quite another alternative to VLIW that is already *sprouting profusely* is multi-core CPUs. Both Intel and AMD have been shipping dual-core chips for quite some time now with quad-core chips promised in 2007. Sun is already shipping 8-core UltraSparc T1 chips, while Rapport Inc. and IBM have already announced development of Kilocore technology that allows combining as many as 1,024 8-bit processors with a PowerPC core on a single low-cost chip. Thus extrapolating current trends we are likely to see further profusion of multi-core CPUs from all leading manufacturers, especially for server markets. Chances are that as number of on-chip cores grows the cores themselves would become more simple and less-deeply pipelined (kind of like UltraSparc T1 is doing already). We are also likely to see some dedicated co-processor-like cores suitable for performing **SIMD**/multimedia instructions while other cores might be *deprived* of such capacity *in favor of* improved energy efficiency and increased overall number of cores.

Perhaps the most *noteworthy* point is that we are unlikely to see dramatic single-threaded code performance improvements unless a way of frequency increases is found that does not result in the market increase in power consumption (for example, new manufacturing technology in the vein of IBM's recent report of experimental SiGe chips running at 350 GHz at room temperature and at 500 GHz when *chilled* by liquid *helium*).

And the truth is that there is no *compelling* need for further raw CPU speed increases for the following key reasons:
- Computers are already much more powerful than most common tasks require.
- Code efficiency is at all time low and potentially hide at least an order of *magnitude* performance boost if we just optimize the code.
- Memory and I/O bottlenecks are most common causes of slow-down.

What is amazing is that for a long time we have been using only *a handful of* CPU models under the *aegis* of general-purpose computing. Furthermore we thought that a better CPU makes a better computer, which is no longer so. What seems to be more important now is overall system design rather than just CPU design, and we are likely to see more system and CPU specialization (and models) targeting different application areas.

Emerging Processor Lines

We already see three major lines of processors targeting mobile, desktop, and server markets. This trend is likely to continue and result in appearance of even more processor lines optimized not only for various segments but for various applications or intended uses as well. For instance, for application servers we may see Intel and AMD delivering vastly multi-core CPU with good integer capacity and dedicated encryption/decryption hardware in the *vane* of Sun's UltraSparc, while in mobile market we may see *stripped-down* extra-low-power CPUs that ensure very long batter life, perhaps with finer frequency scaling similar to what *coarse-grained* AMD's PowerNow! technology does now. There certainly seems to be a room for lower-performance CPUs for ultra-mobile computers since most of them are used for reading, browsing and other simple tasks that do not require much of CPU power (specific tasks such as multimedia encoding/decoding and 3D graphics are already partially **offloaded** to dedicated hardware and are likely to be even more **confined** to specialized chips in the future).

So focusing on mobile CPU market it is clear that power efficiency and that not only of CPU but of the entire system is likely to be much more important than raw processor speed. After all most mobile users are not likely to exploit potential CPU performance to the fullest extent unless we throw at them really bad code. Almost commodity pricing on computational power today is such that consumers can afford buying more and more specialized hardware that is better suited for a particular purpose thus fulfilling Bill Gates' vision of computers in every pocket. This is in fact already happening as we all are **grabbing** iPods, cell phones, PDAs, and BlackBerry devices to **complement** our laptops and desktop PCs. No more one-size fits all. This is the most certain prediction that one can make about future CPUs. We shall see more and more specialized models and not necessarily more powerful ones. *Thus as far as mobile market is concerned we might see CPUs with more finely grained frequency control that responds to idle time, variable rotation rate hard drives and possibly stripped out of some advanced features such as enhanced multimedia processing instructions in favor of dedicated hardware performing the latter tasks.*

In fact AMD is already making some steps in this direction with its upcoming 4×4 platform and open specification enabling 3rd party co-processor design. In the long term it makes little sense to **burden** CPU with DVD playback or SSL encryption. These and similar tasks should and with time will be handled completely by dedicated hardware that is going to be far more efficient (power and performance-wise) than CPU. Further variety of coprocessors will allow enhanced physics and environmental effect experience for gaming enthusiasts and improved performance for scientific/multimedia applications. Thus the role of CPU is likely to diminish with time living little reason for further clock-speed improvement.

Frankly the role of CPU as a *jack* of all trades started to *wane* with the advent of GPUs. 3D graphics was the most compelling reason to boost CPU power. Now PCs typically have a dedicated processor (or two in the case of AMD's 4×4 platform) that is far better suited for the task. Similarly most music/multimedia hardware relies on its own expansion boards *outfitted* with custom logic/DSP processors (take ProTools or Creamware products, for example). And with time we are likely to end up with a motherboard design that would contain numerous specialized chips or co-processors designed with a single task in mind. So in this respect we are back to the single-purpose computing we have *started with*, although such return is a mere new loop in the spiral.

Ironically, return to special-purpose computing results in further relaxing of requirements for higher processor performance: Special-purpose code is usually better optimized and thus can perform equally well on much slower CPUs. In reality most hand-held devices are powered with few hundred MHz CPUs that are capable of providing similar experience (save for small screen and tight keypad) we have with our gigahertz-fast desktop PCs. Similarly specially-designed DSPs are far better for **MPEG** playback or sound processing than general-purpose CPU that can do the same running at high GHz.

In other words, what is likely to happen is that CPU frequency increases will become very **modest** in the near future. As hardware manufacturers compete for the markets we are likely to see less and less general-purpose and more and more specialized hardware for various purposes. Perhaps in 10 years today's Athlon and Xeon CPUs would seem like *dinosaurs*, hot, big, and less than bright, with the role of CPU in the computer reduced from the do-it-all-yourself to coordinate-the-work-of-others.

Conclusion

There is another compelling reason to believe that big, hot, and *insanely* fast CPUs will die out due to natural selection. *As people become more and more aware of "green" concepts and conscious of power consumption our eyes will finally open to extremely bloated code that* **out** *GHz-rated CPUs execute at the same rate as MHz-rated processors in specialized devices.* The proper question to ask would be "How much power does your software require?", where power means electricity with the implication of the high energy cost. Indeed that would mean that slow and bloated software is expensive software for it requires CPU to run at *full blast*. To make this point more clear think of a datacenter with a thousand blade servers with each server sporting several CPUs and hard disks. Bloated and slow software that we have today implies that the operating cost of the datacenter is high for it needs a thousand blade servers, thousand terabyte disks, and gigabytes and gigabytes of memory with cooling and power cost of 10 million a year. Now let's say if we are to optimize our software to reduce RAM, disk, and CPU performance requirements by an order of magnitude (which is easily achieved if we scrap interpreted and otherwise "managed" code with inefficient memory management model and multi-layered libraries and invest in compiler and optimizer development) and reduce the number of servers 10 times? Or instead replacing huge blade servers with gigahertz CPUs with compact pocket-size microblades outfitted with megahertz-rated CPUs, few megabytes of RAM and a microdrive?

Needless to say, there is *amble* room for software optimization that has been ignored for decades since the increases in CPU performance allowed us to neglect it. Yet now the situation with energy resources is such that slow and bloated software means higher costs both in direct electric power required by CPU to process it and indirectly in power consumed by RAM, enormous hard drives and *cumulative* cooling costs. Furthermore recent tendency to aggregate multiple software components running on shared computational resource (that is, a server) under control of a multitasking OS should be reversed in favor of completely isolated software components running on low-power dedicated hardware. Thus if we are to begin optimizing our code we are likely to see blade server *racks* replaced with microblade server racks where each microblade is performing a dedicated task, consuming less power; and where the total number of microblades is much greater than the number of initial "macro" blades.

Indeed such complete isolation of software components (database instances, Web applications, network services, and the like) that are currently squeezed together on the same server should greatly improve system robustness due to the possibility of real-time component hot-swap or upgrade and completely eliminating software installation, deployment and patch conflicts that **plague** *large servers of today.*

When and whether that will happen depends on two factors energy costs and code optimization efficiency. The former drives the latter. Therefore further increase in energy prices is likely to result in gradual reduction of the role of CPU in computer system, more optimized code and return towards single-processor/single-task special-purpose computing paradigm. On the other hand this vision may never *materialize* if a technological breakthrough occurs on manufacturing side that would allow further CPU speed increases without the increased energy dissipation (quantum computing, advances in *superconductors*, *photonics*, and so on). However, one thing is clear – the role of CPU performance is definitely waning, and if a radical new technology fails to materialize quickly we will be compelled to write more efficient code for power consumption costs and reasons.

Words

hone[həun] v.使完美，使更有效	**grab**[græb] v.抓取，赶
flop[flɔp] v.彻底失败，砸锅	**complement**['kɔmplimənt] v.补助，补足
tarnish['tɑ:niʃ] v.减损，使…成为泡影	**burden**['bə:dn] v.负担
prospect['prɔspekt] n.前景，机会	**jack**[dʒæk] n.[俚]火车头
niche[nitʃ] n.合适的环境，产品或服务所需的特殊领域	**wane**[wein] v.衰落
sprout[spraut] v.萌芽，迅速成长和出现	**outfit**['autfit] v.配备，装备
profusely[prəu'fju:sli] adv.丰富地	**modest**['mɔdist] v.适度的，适中的
deprive[di'praiv] v.剥夺，使丧失	**dinosaur**['dainəsɔ:] n.庞然大物，恐龙
noteworthy['nəutwə:ði] adj.显著的，引人注意的	**insanely**[in'seinli] adv.疯狂地，狂暴地
chill[tʃil] v.降低…的温度，使冷却	**out**[aut] adj.过时的
helium['hi:ljəm] n.氦（化学元素）	**amble**['æmbl] n.缓步走动
compelling[kəm'peliŋ] adj.引人注目的	**cumulative**['kju:mjulətiv] adj.累积的
magnitude['mægnitju:d] n.量级	**rack**[ræk] n.支架
aegis['i:dʒis] n.支持	**plague**[pleig] v.困扰
vane[vein] n.风向标	**materialize**[mə'tiəriəlaiz] v.使成真，实现
offload[ɔf'ləud] v.卸载	**superconductor**[,sju:pəkən'dʌktə] n.超导体
confine['kɔnfain] v.限制，使局限于	**photonics**[fəu'tɔniks] n.光子学
	radical['rædikəl] adj.根本的，基础的

Phrases

in favor of	支持，有利于	**start with**	首先，第一
a handful of	少数	**full blast**	最大限度地
strip-down	脱去，拆开	**hot-swap**	热插拔
coarse-grained	质地粗糙的		

Abbreviations

SIMD	Single Instruction Multiple Data	单指令多数据流
MPEG	Moving Pictures Experts Group	动态图像专家组，视频、音频、数据的压缩标准

Complex Sentences

1. **Original:** Yet the state of current technology is such that current VLIW/EPIC compilers are not yet good enough for general purposes and therefore theoretically possible performance gains are almost never achieved (VLIW processors can execute as many as 32 instructions in parallel if a

compiler can find and schedule that many).

Translation: 然而，当前的技术现状是这样的：目前的VLIW/EPIC编译器尚未足够满足通用的需求，因此理论上可能的性能提高几乎不可能达到（倘若一个编译器能够找到并调度这么多指令的话，VLIW处理器能够并行执行多达32条指令）。

2. **Original:** Thus as far as mobile market is concerned we might see CPUs with more finely grained frequency control that responds to idle time, variable rotation rate hard drives and possibly stripped out of some advanced features such as enhanced multimedia processing instructions in favor of dedicated hardware performing the latter tasks.

Translation: 因此就移动市场而言，我们可能会看到这样的CPU，它具有响应空闲时间的更细粒度的频率控制，以及可变转速的硬盘驱动器，并可能剥离一些高级功能，例如：支持专用硬件执行后面的任务的增强型的多媒体处理指令。

Exercises

I. Read the following statements carefully, and decide whether they are true (T) or false (F) according to the text.

____1. For a long time we had thought that a better CPU makes a better computer.

____2. IBM's SiGe chips present a way of frequency increases that does not result in increasing power consumption.

____3. We may boost performance at least by an order of magnitude just by optimizing the code.

____4. We are likely to return to the special-purpose computing in a new loop in the spiral.

____5. Hardware that is more dedicated and efficient than CPU is better suited for various complex tasks.

II. Choose the best answer to each of the following questions.

1. Which of the following is WRONG about VLIW?

(A) VLIW failed on general purpose due to the lack of compilers, cross-compilers and automatic code optimization techniques.

(B) Commercial VLIW failure has disappointed involved companies and resulted in their ceasing attempt.

(C) In contrast to VLIW, a multi-core CPU is already booming as another alternative to it.

(D) VLIW may find its niche in markets for solving a very limited set of special-purpose tasks.

2. Which of the following is WRONG about the trend of CPUs?

(A) Increasing overall number of on-chip cores

(B) Lower-performance for ultra-mobile computers

(C) Increasing raw processor frequency

(D) Improved power efficiency

3. Which of the following does seem the most feasible about the future of computing?

(A) Restoring CPU to the role of a jack

(B) Increasing raw CPU speed further

(C) Desiring quick technological breakthrough on energy saving

(D) Optimizing code and software

III. Translating.

1. As people become more and more aware of "green" concepts and conscious of power consumption our eyes will finally open to extremely bloated code that out GHz-rated CPUs execute at the same rate as MHz-rated processors in specialized devices.

2. Indeed such complete isolation of software components (database instances, Web applications, network services, and the like) that are currently squeezed together on the same server should greatly improve system robustness due to the possibility of real-time component hot-swap or upgrade and completely eliminating software installation, deployment and patch conflicts that plague large servers of today.

Part 2　Simulated Writing: Cover Letter

Introduction

When applying for a job, you usually need to submit both a resume and a cover letter. A cover letter markets your skills, abilities, and knowledge. It must be persuasive and accomplish four tasks:

- Catch the reader's attention
- Explain which particular job interests you and why
- Convince the reader that you are qualified for the job by drawing your reader's attention to particular elements in your resume
- Request an interview

Structure

1. Opening Paragraph

Start with an interest-creating sentence in which you appeal to the employer's needs. To catch the reader's attention, the opening paragraph must be favorable and concise:

- Indicate how you heard about the opening. If you have been referred to the job opportunity by an employee of the company or someone else, be sure to mention this even before you state your job objective. This catches the reader's attention indicating you are familiar with people in the company.

- State your job objective and mention the specific job title.
- Explain why you are interested in the job.

2. Middle Paragraph

In the second and third paragraphs, show through examples that you are highly qualified for the job. Limit each of these paragraphs to just one basic point that is clearly stated in the topic sentence. For example, the second paragraph might focus on work experience and the third paragraph on educational achievements. Don't just tell readers that you're qualified – show them by including examples and details. Indicate how (with your talents) you can make valuable contributions to their company. At the end of this paragraph make a reference to your resume.

3. Closing Paragraph

In the final paragraph, request an interview. Let the reader know how to reach you by including your phone number or email address. End with a statement of good will, even if it is only "thank you".

Sample

P.O. Box 112, Beihang University
Beijing 100191, PR China

January 10, 2009

Ms. Patricia White
Manager, Information System Department
ABC Software & Services, Inc.
World Trade Center
7 South Avenue, Chaoyang District
Beijing 100004, PRChina

Dear Ms. White,

I am writing to express my interest in the software engineer position advertised in the January 3rd edition of ABC Recruitment Website. I will graduate from Beihang University in July with a Bachelor of Engineering degree in software engineering.

You indicated a need for someone with both engineering experience and English communication skills. During my junior and senior years in Beihang University, I have worked as a student designer and programmer in the College Information System Office and the Digital Earth & GIS Lab. In addition, I have been a teaching assistant in the college for two years; and as an intern market research analyst in FE Software, Inc., I worked in a complete English environment for 10 months. Please see my enclosed resume for a more complete view of my background.

I am eager to learn more with this opportunity about ABC Software & Services, Inc., and look forward to interviewing with you at your convenience. I can be reached by phone at (010) 1234-5678 or by e-mail at mark@bh.edu.cn. Thank you for your time and consideration.

Sincerely,

Mark Li

Enclosure: Resume

Part 3　Listening and Speaking

≫≫ Dialogue: Interview

(Mark's application for a software engineer position in ABC Software & Services, Inc. has been responded to by the Information System Department of the company. Today he comes to the company for an interview.)

Ms. White	Good morning, thank you for applying for our opening positions. My name is Patricia White, manager of the Information System Department.
Mark	Good morning, Ms. White. My name is Mark Li, and I'm coming for a software engineer position in the Information System Department.
Ms. White	Well, now I would like to talk with you regarding your qualification for this interview. What can you tell me about yourself ?[1]
Mark	I'm an active and ***earnest*** person and like to work in a team. I like to communicate with people. I'm very interested in the latest advancements in science and technology. I love the IT industry and want to pursue an IT career.

Ms. White	Why did you apply for a job at our company?[2]
Mark	As we know, ABC Software & Services, Inc. is not only one of the leading software products and services providers in China, but also has a global **presence**. So I believe that working here can help me grow professionally. I also think I can contribute to the development of the company.
Ms. White	Have you ever done an **internship** that helped to prepare you for this type of work?[3]
Mark	Yes. During my junior and senior years in Beihang University, I have worked as a student designer and programmer in the College Information System Office and the Digital Earth & GIS Lab, having participated in several practical software projects including a platform of Digital Earth **sponsored** by National Fund, a 3D simulation system for airplane, Resource Management System for the College Financial Office and Information Management System for the College Library. In addition, as an **intern** market research analyst in FE Software, Inc. I worked in a complete English environment for 10 months. All of the above are very valuable internship experience for me to prepare for this work, I think.
Ms. White	OK, that sounds great! What else do you want to know about your work?
Mark	I want to know what other qualifications you consider necessary for this position?
Ms. White	Well, as far as I know, sometimes the workload of developing software may be really very heavy, so the person who is *in charge of* it may have to be under much pressure and must be energetic and capable.
Mark	Yes, I agree with you. On this point, my internship experience has helped me to get used to this kind of work environment
Ms. White	OK, Thank you for your interest in our company. If we decide to bring you onto our team, you will receive an e-mail from us within five working days.
Mark	Thank you very much. I look forward to hearing from you.

Exercises

Work in pairs, and make up a similar conversation by replacing the scenarios with other material in the below.

➡ Interview (Another)

[1] About self-introduction
  1. Tell me something about yourself.
  2. How do you *value*/judge yourself?
  3. What is your work style?
  4. Why do you think that you deserve to get the job?

[2] About applying motivation
  1. Why do you want to work for us?
  2. What do you know about us?

[3] About career related experience
  1. How have you prepared yourself for the transition from college to the workplace?
  2. Did you *get any hands on* experience in College?
  3. What aspects of your abilities and experience make you think that you will succeed in this job?

 Words

earnest[ˈə:nist] *adj.*认真的，有决心的	**sponsor**[ˈspɔnsə] *v.*资助，赞助
presence[ˈprezns] *n.*势力	**intern**[inˈtə:n] *n.*实习生
internship[ˈintə:nʃip] *n.*实习（职位）	**value**[ˈvælju:] *v.*评价

 Phrases

in charge of	负责，管理	**get one's hands on**	把…弄到手

⟫⟫ Listening Comprehension: Quantum Computer

Listen to the passage and the following 3 questions based on it. After you hear a question, there will be a break of 10 seconds. During the break, you will decide which one is the best answer among the four choices marked (A), (B), (C) and (D).

Questions

1. What is the meaning of "*a blend or superposition of these classical states*"?
2. What kind of problems does quantum computer particularly fit for?
3. What is the prospect of quantum computer according to this article?

Choices

1. (A) A qubit can exist as 0 or 1 alternatively

   (B) A qubit can exist as both 0 and 1 simultaneously

   (C) A qubit can exist as 0 or 1 in multiple bits

   (D) A qubit can exist as 0 or 1 in a classical bit

2. (A) Problems about physical phenomenon in nature

   (B) Problems with three different states

   (C) Problems with a large amount of data or variables

   (D) Problems with not binary but rather more quaternary

3. (A) Negative   (B) Affirmative   (C) Indifferent   (D) Paradoxical

 Words

staple['steipl] *n.*主要产品，日常必需品	**coefficient**[kəui'fiʃənt] *n.*系数
cage[keidʒ] *v.*将···放入，将···限制于	**uncrackable**[ʌn'krækəbl] *adj.*不可破解的
qubit['kju:bit] *n.*量子比特	**ciphers**['saifə] *n.*密码
quaternary[kwə'tə:nəri] *adj.*四进制的	**infancy**['infənsi] *n.*初期
superposition[.sju:pəpə'ziʃən] *n.*重叠	**cryptography**[krip'tɔgrəfi] *n.*密码使用法，密码系统

 Phrases

come of	由···引起

⟫ Dictation: Native XML Database

This passage will be played THREE times. Listen carefully, and fill in the blanks with the words you have heard.

When your only tool is a hammer, everything looks like a nail. When your only tool is a relational database, everything looks like a table. ____1____ , however, is more complicated than that. Data often isn't *tabular* and can benefit from a tool that more closely fits its ____2____ structure.

A great deal of data is now being ____3____ in XML, and more is being created every day.

However, relational databases don't really fit XML documents, at least not in their full____4____.
While you **shred** an XML document to **stuff** it into a relational table, it tends to lead to the ___5___
of details like element order, processing instructions, ____6____ , and other elements that are
important in many ____7____ in which XML documents don't look____8____ like serialized
tables in the first place.

The ____9____ tool for managing XML document might well be native XML databases. A
native XML database is one that treats XML documents and elements as the fundamental___10___
rather than tables, records, and fields. It defines a logical model for an XML document—as
opposed to the data in that document—and stores and ____11____ documents according to that
model. The logical model of ____12____ documents is called "collections" in general.
Databases can set up and manage many collections at one time. In some implementations, a
hierarchy of collections can exist, much in the same way that an operating system's____13____
structure works. On the other hand, a native XML database need not have any particular ___14___
physical storage model. For example, it can be built on a relational, hierarchical, or object-oriented
database, or use a proprietary storage format such as ____15____,compressed files.

Such a database keeps all your content in one easily searched, easily managed place. All you
have to do to get the information out is to make a ____16____ . It allows retrieving the original
unparsed document, character-per-character or even byte-per-byte; and enables developers to use
tools and languages that more naturally fit the structure of the documents they're working with,
thereby enhancing ____17____ .

Additionally, queries over a well-designed, well-implemented native XML database are
simply faster than queries over documents stored in a ____18____ system. It is also widely
believed that native XML databases can ____19____ **outperform** traditional relational databases
for tasks that involve ____20____ document processing, such as newspaper publishing, Website
management, and Web services.

Words

tabular['tæbjulə] *adj.*表格式的`	**stuff**[stʌf] *v.*塞满，填充
shred['ʃred] *v.*切碎	**outperform**[autpə'fɔ:m] *v.*优过，胜于

Glossary

A

a bulk of 大部分

a handful of 少数

abbreviate[ə'bri:vieit] v.缩写，简写

abnormal[æb'nɔ:məl] adj.反常的，异常的

above all 最重要的是，首先

abridge[ə'bridʒ] v.删节，精简

abstraction[æb'strækʃən] n.抽象

accommodate[ə'kɔmədeit] v.和解，调和

account for 说明，解释

acknowledged[ək'nɔlidʒid] adj.公认的，被普遍认可的

act[ækt] n.节目

adapter[ə'dæptə] n.适配器

address[ə'dres] v.处理，对付

ad-hoc['æd'hɔk] adj.特别的，特定的

advance[əd'va:ns] v.促进，建议

advent['ædvənt] n.出现，到来

advocate['ædvəkit] v.提倡，主张

adware['ædwɛə] n.广告软件

aegis['i:dʒis] n.支持

aerial['ɛəriəl] adj.航空的，由飞机进行的

afar[ə'fa:] adv.遥远地

aggregation[ægri'geiʃən] n.聚合

agile['ædʒail] adj.敏捷的，灵活的

albeit[ɔ:l'bi:it] conj.虽然

algorithm['ælgəriðəm] n.算法

all along 连续，始终

all of a sudden 突然

allegedly[ə'ledʒidli] adj.声称的

allude[ə'lju:d] v.暗指，间接提到

amateur['æmətə] n.业余爱好者

amble['æmbl] n.缓步走动

amicably['æmikəbli] adv.友善地

amigo[ə'mi:gəu] n.朋友

among others[other things] 其中

amongst others 其中

amount to 总计

analogous[ə'næləgəs] adj.类似的，相似的

anew[ə'nju:] adv.重新，再

animation[,æni'meiʃən] n.动画片，卡通

answer['a:nsə] v.满足，适用

anticipate[æn'tisipeit] v.预期，预见

antitrust[,ænti'trʌst] adj.反托拉斯的，反垄断的

anti-virus 防（计算机）病毒

applet['eplet] n.Java小应用程序，application let

appliance[ə'plaiəns] n.用具，装置

apt[æpt] adj.易于…的

arcade[a:'keid] n.游乐中心，娱乐厅

arcane[a:'kein] adj.晦涩难解的，神秘的

architect['a:kitekt] v.设计，建造

article['a:tikl] n.商品，项目

artifact['a:tifækt] n.人工制品

artist['a:tist] n.骗子，家伙

as far as 直到，远到

as of 到…为止

as with 正如…一样

as yet 到目前为止

ask for 请求，寻找

aspect['æspekt] n.方面

assembler[ə'semblə] n.汇编程序

assembly[ə'sembli] n.汇编，集合

assembly language 汇编语言

assert[ə'sə:t] v.断言，声称

assimilate[ə'simileit] v.吸收

astonish[əs'tɔniʃ] v.使惊讶

asynchronous[ei'siŋkrənəs] adj.不同时的，异步的

at hand 在手边，在近处

at large 随便地，笼统地

atomic[ə'tɔmik] *adj.*原子的

augment[ɔ:g'ment] *v.*扩大，增加

authentic[ɔ:'θentik] *adj.*可信的，可靠的

author['ɔ:θə] *v.*制作，创作

authorization[,ɔ:θərai'zeiʃ nə] *n.*授权，认可

automata[ɔ:'təmətə] *n.*自动机

autonomy[ɔ:'tɔnəmi] *n.*自治权，自主权

avionics[,eivi'ɔniks] *n.*航空电子技术

B

back up 备份

backbone['bækbəun] *n.*主干网，广域网中的一种高速链路

back-end 后端

backlink[bækliŋk] *n.*反向链接

back-office 后台

backplane[bækplein] *n.*底板

balance['bæləns] *n.*收支差额，结余

bandwidth['bændwidθ] *n.*带宽

be acquainted with 了解，结识

be attentive to 注意，留心

be charged with 负···责任

be concerned with 参与

be coupled with 和···联合，结合

be faced with 面临

be subject to 附属于，易受···的

be suited to 适合于

beep[bi:p] *n.*嘟嘟声

benchmark['bentʃ,ma:k] *n.*基准

beneficiary[beni'fiʃ əri] *n.*受惠者，受益人

beware[bi'wɛə] *v.*小心，谨防

bidder['bidə] *n.*出价人，投标人

bidirectional[,baidi'rekʃ ənəl] *adj.*双向的

bill[bil] *v.*宣布，公告

binary['bainəri] *n.*二进制

bind[baind] *v.*使受法律（或合同、道义等的）约束

biometric[,baiəu'metrik] *adj.*生物鉴别法的

blade[bleid] *n.*刀片

bloat[bləut] *v.*膨胀

block[blɔk] *n.*区域

blog[blɔg] *n.*博客

blogger['blɔgə] *n.*写博客的人

blow[bləu] *v.*烧断，熔断

bluetooth['blu:tu:θ] *n.*蓝牙

boast[bəust] *v.*以有···而自豪

boilerplate['bɔiləpleit] *n.*样板文件

boom[bu:m] *v.*兴隆，繁荣

boot[bu:t] *v.*引导

boot sector 引导扇区

bootable['bu:təbl] *adj.*可引导的

bootstrap['bu:tstræp] *v.*引导

bot[bɔts] *n.*机器人程序

branch[brɑ:ntʃ] *n.*分支程序

brand name computer 品牌计算机

branding[brændiŋ] *n.*商标，品牌

breach[bri:tʃ] *n.*破坏

break[breik] *v.*暂停工作

break into 闯进

break up 分解

breakthrough['breik'θru:] *n.*突破，重大成就

breathe[bri:ð] *v.*将···注入

breed[bri:d] *n.*品种，种类

brilliant['briljənt] *adj.*超群的，杰出的

broadband['brɔ:dbænd] *n.*宽带

brown-out['braunaut] *n.*部分灯火管制

browser[brauzə] *n.*浏览器

bruise[bru:z] *v.*撞伤，碰伤

bubble['bʌbl] *n.*泡沫，幻想的计划

budget['bʌdʒit] *adj.*合算的，廉价的

buffer overflow 缓冲区溢出

bug[bʌg] *n.*程序缺陷，错误

build[bild] *n.*构造

built-in['bilt'in] *adj.*内置的

burden['bə:dn] *v.*负担

bus[bʌs] *n.*总线

bus snoop 总线监听

bust[bʌst] *v.*破产，失败

button['bʌtn] *n.*按钮

buy out 买下全部

buzz word 热门词语

by analogy with 由…类推
by right of 由于，因为
by the way 顺便提一句
by way of 经由，作为
bypass['baipɑ:s] v.忽视，绕过，回避
bytecode['baitkəud] adj.字节码

C

cache[kæʃ] n.高速缓存
cache snoop 高速缓存监听
cage[keidʒ] v.将…放入，将…限制于
call for 要求，提倡
camcorder['kæmkɔ:də] n.可携式摄像机
canonical[kə'nɔnikəl] adj.规范的
capacitance[kə'pæsitəns] n.电容，电容量
capacitor[kə'pæsitə] n.电容器
Capitol Hill 美国国会山，美国国会
cargo['kɑ:gəu] n.货物
carrier['kæriə] n.载波（信号）
case[keis] n.案例，实例
cater['keitə] v.满足（需要），投合
caution['kɔ:ʃən] v.警告
caveat['keiviæt] n.警告
census['sensəs] n.人口普查
certify['sə:tifai] v.证明，确认
chargeback['tʃɑ:dʒbæk] n.退款
check[tʃek] n.支票
check out 结账
checkout['tʃekaut] n.结算
chihuahua[tʃi'wɑ:wɑ:] n.吉娃娃（一种产于墨西哥的狗）
chill[tʃil] v.降低…的温度，使冷却
chimera[kai'miərə] n.嫁接杂种
chip[tʃip] n.芯片
chipset[tʃipset] n.芯片集
chore[tʃɔ:] n.日常工作，例行事务
CID 一种浏览器病毒
ciphers['saifə] n.密码
circa['sə:kə] adv.大约
circumstance['sə:kəmstəns] n.环境，情况
claim[kleim] n.索赔，赔款
clandestine[klæn'destin] adj.秘密的

class[klɑ:s] n.类
client['klaiənt] n.客户端
cloak[kləuk] v.掩饰，掩盖
clone[kləun] v.复制，克隆
cluster['klʌstə] n.集群
coarse-grained 质地粗糙的，纹理粗的
coaxial[kəu'æksəl] adj.同轴的
codebase['kəud'beis] n.代码库
code-name[kəud'neim] v.给与代号
coefficient[kəui'fiʃənt] n.系数
coherent[kəu'hiərənt] adj.一致的，连贯的
cohesive[kəu'hi:siv] adj.黏着的
coin[kɔin] v.杜撰，设计
coincide[,kəuin'said] v.同时发生
collie['kɔli] n.牧羊犬（一种源于苏格兰的高大聪敏长毛的牧羊狗）
column['kɔləm] n.专栏
come about 发生
come across 偶然遇见
come into action 起作用，投入战斗
come of 由…引起
come up with 提出
commensurate[kə'menʃərit] adj.成比例的，相称的
commitment[kə'mitmənt] n.托付，交托，致力
commodity[kə'mɔditi] n.日用品，商品
compatibility[kəm,pæti'biliti] n.兼容
compatible[kəm'pætəbl] adj.兼容的
compelling[kəm'peliŋ] adj.引人注目的
compiler[kəm'pailə] n.编译器
complement['kɔmplimənt] v.补助，补足
comply[kəm'plai] v.遵从，遵照
comprehensively[,kɔmpri'hensivli] adv.全面地
compromise['kɔmprəmaiz] v.妥协，折衷，危及…的安全
conceal[kən'si:l] v.隐藏
conceive[kən'si:v] v.构思，设计
concern[kən'sə:n] n.担心，忧虑
concurrent[kən'kʌrənt] adj.并发的
confidentiality[,kɔnfi,denʃi'æliti] n.机密性
configuration[kən,figju'reiʃən] n.配置
configure[kən'figə] v.配置

confine['kɔnfain] v.限制，使局限于

confound[kən'faund] v.使困惑，混乱，混淆

conjecture[kən'dʒektʃə] v.推测，猜想

console['kɔnsəu] n.控制台，操纵台

constraint[kən'streint] n.约束，强制

contingency[kən'tindʒənsi] n.意外事故，偶然，可能性

coordinates[kəu'ɔ:dinit] n.坐标

copyleft['kɔpileft] v.非盈利版权

cordless['kɔ:dlis] adj.无绳的

cornerstone['kɔ:nəstəun] n.基础

corruption[kə'rʌpʃən] n.变坏

couch[kautʃ] n.睡椅，沙发

count as　认为…，当作…

counter['kauntə] v.反对，辩驳

countermeasure['kauntə,meʒə] n.对策，反措施

course[kɔ:s] n.航道

crack[kræk] v.解开（秘密等）

cracker['krækə] n.骇客

cram[kræm] v.填满

credit card　信用卡

creep[kri:p] n.蠕变

crimped[krimpt] adj.起皱褶的，有波纹的

cripple['kripl] v.削弱

crunch[krʌntʃ] v.运行，处理

cryptographic[,kriptə'græfik] adj.密码的，暗号的

cryptography[krip'tɔgrəfi] n.密码使用法，密码系统

culminate['kʌlmineit] v.达到顶点，告终

culmination[kʌlmi'neiʃ(ə)n] n.顶点

cultivate['kʌltiveit] v.培养，养成

cumulative['kju:mjulətiv] adj.累积的

curb[kə:b] v.抑制

cure[kjuə] v.改正，消除，治疗

curiosity[,kjuəri'ɔsiti] n.好奇心

curly-brace　卷曲的大括号

custom['kʌstəm] adj.定做的，定制的

cutting edge　尖端

cycle-stealing　周期侵占

D

daemon['di:mən] n.Unix和其他多任务处理操作系统中一种在后台运行的计算机程序

DaimlerChrysler['daimlə'kraislə] n.戴姆勒-克莱斯勒

data mining　数据挖掘

data warehouse　数据仓库

database['deitəbeis] n.数据库

deadlock['dedlɔk] n.死锁

deal with　与…打交道（做买卖）

debit card　借记卡

debug[di:'bʌg] v.调试

deceptively[di'septivli] adv.欺骗地，虚伪地

declarative[di'klærətiv] adj.说明的，陈述的，公布的

decomposition[,di:kɔmpə'ziʃən] n.分解

decouple[di'kʌpl] v.分离，消除…间相互影响

dedicate['dedikeit] v.奉献，致力

dedicated['dedikeitid] adj.专用的

deem[di:m] v.认为

defacto[di:'fæktəu] adj.事实上的，实际上的

default[di'fɔ:lt] n.默认（值），缺省（值）

defragmenter[di:frægməntə] n.磁盘碎片整理程序

defragmenting[di:'frægməntiŋ] n.磁盘碎片整理程序

delegate['deligit] n.委托

deliberate[di'libəreit] adj.故意的，蓄意的

deliberately[di'libərətli] adv.故意地

delve[delv] v.挖掘

demographic[di:mə'græfik] adj.人口统计学的

deprive[di'praiv] v.剥夺，使丧失

derivative[di'rivətiv] n.衍生物，派生的事物

descendant[di'send(ə)nt] n.子孙，后代

deserve[di'zə:v] v.应受，应得

designate['dezigneit] v.选定，指派

desirable[di'zaiərəbl] adj.值得做的，值得要的，有利的

desktop microcomputer　台式微型计算机

detour['di:tuə] n.迂回，弯路

devise[di'vaiz] v.设计

dial-up　拨号（上网）

diaper['daiəpə] n.尿布

die[dai] n.管芯

dinosaur['dainəsɔ:] n.庞然大物，恐龙

discard[dis'kɑ:d] v.丢弃，抛弃

discipline['disiplin] v.通过教学和实践训练

discrepancy[disˈkrepənsi] *n.*差异，矛盾

discrete[disˈkriːt] *adj.*分离的，离散的

discrete graphics card 独立显卡

disjoint[disˈdʒɔint] *adj.*不相交的，没有交集的

disk drive 磁盘驱动器

disparate[ˈdispərit] *adj.*异类的，完全不同的

dispenser[disˈpensə] *n.*自动售货机，自动发放器

dispute[disˈpjuːt] *n.*纠纷，争端

disrupt[disˈrʌpt] *v.*干扰，扰乱，使中断

disseminate[diˈsemineit] *v.*散布

dissemination[diˌsemiˈneiʃən] *n.*分发

dissipate[ˈdisipeit] *v.*驱散

distinguish[disˈtiŋgwiʃ] *v.*使杰出，使著名

distortion[disˈtɔːʃən] *n.*变形，失真

distribution[ˌdistriˈbjuːʃən] *n.*销售版本，销售形式

document[ˈdɔkjumənt] *v.*证明，评述

domain[dəuˈmein] *n.*域名

dominance[ˈdɔminəns] *n.*优势，统治

downgrade[ˈdaungreid] *n.*向下渐变

downplay[ˈdaunplei] *v.*不予重视

dramatically[drəˈmætikəli] *adv.*显著地，引人注目地

drastically[ˈdræstikəli] *adv.*激烈地，彻底地

draw on 引起，利用

drive[draiv] *n.*驱动器

driver[ˈdraivə] *n.*驱动器，驱动程序

drop out of 退出

dropper[ˈdrɔpə] *n.*落下的人或物

dub[dʌb] *v.*授予称号

dumb[dʌm] *adj.*哑的，不智能的，被动的

duplex[ˈdjuːpleks] *adj.*双工的

duplicable[ˈdjuːplikəbl] *adj.*可复制的，可再发生的

E

earnest[ˈəːnist] *adj.*认真的，有决心的

ease of 解除，减少

eavesdropper[ˈiːvzˌdrɔpə] *n.*偷听者

edge[edʒ] *n.*边缘

effort[ˈefət] *n.*成果，努力的结果

electrolyte[iˈlektrəulait] *n.*电解质，电解液

elegant[ˈeligənt] *adj.*简洁的，典雅的

elicit[iˈlisit] *v.*得出

eligible[ˈelidʒəbl] *adj.*符合条件的，合格的

elite[eiˈliːt] *n.*精锐，杰出人物

embark[imˈbɑːk] *v.*开始，从事

embedded computer 嵌入式计算机

embody[imˈbɔdi] *v.*包含，体现

embrace[imˈbreis] *v.*包括，利用

empirical[emˈpirikəl] *adj.*完全根据经验的，经验主义的

encapsulation[inˌkæpsjuˈleiʃən] *n.*封装

encase[inˈkeis] *v.*嵌入，装入

enchantress[inˈtʃɑːntris] *n.*女巫

encompass[inˈkʌmpəs] *v.*构成，包括

encrypt[inˈkript] *v.*加密

encryption[inˈkripʃən] *n.*加密术，密码术

end up 告终，结束

enforce[inˈfɔːs] *v.*实施，执行

enforceable[inˈfɔːsəbl] *adj.*可实施的，可执行的

engage[ˈstʌmbl] *v.*使卷入其中，与…交战

engaging[inˈgeidʒiŋ] *adj.*有吸引力的，吸引人的

ensue[inˈsjuː] *v.*相继发生

enthusiast[inˈθjuːziæst] *n.*热心家，狂热者

enumeration[iˌnjuːməˈreiʃən] *n.*枚举

enzyme[ˈenzaim] *n.*酶

epithet[ˈepiθet] *n.*绰号，称号，表述词语

ergonomics[ˌəːgəuˈnɔmiks] *n.*人类工程学，生物工程学

erratic[iˈrætik] *adj.*奇怪的

escalate[ˈeskəleit] *v.*逐步增强

espionage[ˈespiənidʒ] *n.*间谍活动

ether[ˈiːθə] *n.*以太

Ethernet[ˈiːθənet] *n.*以太网

euphoria[juːˈfɔːriə] *n.*兴高采烈

every so often 时常，不时

exacerbate[eksˈæsəbeit] *v.*恶化，加剧

exemplar[igˈzemplə] *n.*样本，示例

exhale[eksˈheil] *v.*呼气

exhaustive[igˈzɔːstiv] *adj.*无遗漏的，详尽的

expansion card 扩展卡

explicitly[iksˈplisitli] *adv.*清楚地，明晰地

explode[iksˈpləud] *v.*猛增

exploit[iksˈplɔit] *n.*业绩，功绩

exponentially[ˌekspəu'nenʃəli] v.指数地

extrapolation[ˌekstrəpəu'leiʃən] n.外推法，推断

exuberance[ig'zju:bərəns] n.生气勃勃

F

facade[fə'sɑ:d] n.（房屋的）正面，立面

facilitation[fəˌsili'teiʃən] n.简易化

facility[fə'siliti] n.工具，便利

fade[feid] v.逐渐消失，衰弱

fail-over 故障备份替换

fall to 着手开始工作

familiarize[fə'miljəraiz] v.使熟悉，使被认知

fantasy['fæntəsi] n.幻想，狂想，想象

far-fetched['fɑ:'fetʃt] adj.牵强的，不自然的

fatal['feitl] adj.致命的，毁灭性的

feasibility study 可行性研究

feed[fi:d] n.进给

fend off 挡开，躲开

fiber['faibə] n.纤程

fiddle around 不经意地干活，闲荡

file[fail] v.提出（申请等），呈请把…备案

fine-grain （影像）有微粒的，细致的

fine-tune 调整，微调

firewall['faiəcwɔ:l] n.防火墙

firmware['fə:mˌwɛə] n. 固件（软件硬件相结合）

fixed-line 固定电话

flash drive 闪存

fledgling['fledʒliŋ] adj.年轻的或无经验的

flop[flɔp] v.彻底失败，砸锅

floppy disk 软盘

flowchart[fləu'tʃɑ:t] n.流程图

fluctuation[ˌflʌktju'eiʃən] n.波动，起伏

folksonomy[fəuk'sɔnəmi] n.分众分类（Floks Taxonomy）

following['fɔləuiŋ] n.崇拜者，追随者

footage['futidʒ] n.连续镜头，电影胶片

for a time 暂时

for short 简称，缩写

forerunner[fɔ:'rʌnə] n.先驱

form[fɔ:m] v.建立，组成

form factor 外形

formal method 形式化方法

forms[fɔ:mz] n. 窗体

formulation[ˌfɔ:mju'leiʃən] n.由…制定的配方

forum['fɔ:rəm] n.论坛

fossil['fɔsl] adj.过时的，陈旧的

foster['fɔstə] v.促进，培养

fragment['frægmənt] n.片段，分段

free of charge 免费

free-ride 搭便车

frill[fril] n.装饰

frustrate[frʌs'treit] v.挫败，使感到灰心

full blast 最大限度地

further['fə:ðə] v.促进，推动

fuse[fju:z] n.保险丝

fuse[fju:z] v.合并，结合一起

fuselage['fju:zilɑ:ʒ] n.飞机机身

G

gadget['gædʒit] n.小配件，机械装置，小工具

garner['gɑ:nə] v.得到，收集

gateway['geitwei] n.网关

general['dʒenərəl] n.（做事的）头儿，高级官员

general purpose 多功能的，多用途的

generative['dʒenərətiv] adj.有生产力的，生成的

genetic[dʒi'netik] adj.遗传学的，基因的

genetics[dʒi'netiks] n.遗传学

genuine['dʒenjuin] adj.真实的，真的

geometry[dʒi'ɔmitri] n.几何，几何形状

get one's hands on 把…弄到手

giant['dʒaiənt] n.巨人，伟人

give rise to 导致，使…发生

given that… 假定，已知

go beyond 超出，超过

grab[græb] v.抓取，赶

Grand Canyon （美）科罗拉多大峡谷

grid[grid] n.网格

grocery['grəusəri] n.杂货店

groupware[gru:pwɛə] n.组件，群件

H

hack[hæk] v.黑客入侵

hacker[ˈhækə] *n.*电脑黑客

hand over 移交，交出

handheld computer 掌上电脑

hard disk 硬盘

header[ˈhedə] *n.*制造（钉头）的工具

headquarter[ˌhedˈkwɔːtə] *v.*以…作总部，设总公司于…

heap[hiːp] *n.*堆

helium[ˈhiːljəm] *n.*氦（化学元素）

heterogeneous[ˌhetərəuˈdʒiːniəs] *adj.*异构的

heuristic[hjuəˈristik] *adj.*启发式的

heuristics[hjuəˈristiks] *n.*启发法

hierarchical[ˌhaiəˈrɑːkikəl] *adj.*分等级的

high-end 高端的

hone[həun] *v.*使完美，使更有效

honey pot 贮蜜罐

host[həust] *v.*做主人招待，托管

host[həust] *n.*主机

hot-swap 热插拔

housing[ˈhauziŋ] *n.*壳

how-to 解释作法的，指南的

hub[hʌb] *n.*网络集线器

hub[hʌb] *n.*（兴趣、活动等的）中心

hybrid[ˈhaibrid] *adj.*混合的

hype[haip] *n.*大肆宣传，大做广告

hyper-[ˈhaipə] *pref.*超出，过于

hypermedia[ˈhaipəmiːdiə] *n.*超媒体

hypothesize[haiˈpɔθisaiz] *v.*假设，猜测

I

icon[ˈaikɔn] *n.*图标

Iliad[ˈiliəd] *n.*《伊利亚特》(古希腊描写特洛伊战争的英雄史诗，相传为荷马所作)

imagery[ˈimidʒəri] *n.*影像

immense[iˈmens] *adj.*极大的，巨大的

impose[imˈpəuz] *v.*把…强加给，强派

in accordance with 依照

in brief 简单地说

in charge of 负责，管理

in concert 一致，一齐

in conjunction with 与…协力

in favor of 支持，有利于

in one shot 立刻，马上

in place 在适当的地位

in proportion as 按…比例，和…相称

inadvertent[ˌinədˈvəːtənt] *adj.*疏忽造成的

incorporate[inˈkɔːpəreit] *v.*合并，加入

indicator[ˈindikeitə] *n.*指针，指示器

inescapable[ˌinisˈkeipəbl] *adj.*不可避免的，不可忽视的

infamous[ˈinfəməs] *adj.*声名狼藉的

infancy[ˈinfənsi] *n.*初期

inflated[inˈfleitid] *adj.*夸张的

inhale[inˈheil] *v.*吸气

inheritance[inˈheritəns] *n.*继承

inline[inlain] *adj.*内嵌的

insanely[inˈseinli] *adv.*疯狂地，狂暴地

inspire[inˈspaiə] *v.*鼓舞，唤起

installation[ˌinstəˈleiʃən] *n.*装置，设备

instance[ˈinstəns] *n.*实例

instruction[inˈstrʌkʃən] *n.*指令

insurmountable[ˌinsəˈmauntəbl] *adj.*不能克服的，不能超越的

integrated circuits 集成电路

integrated graphics card 集成显卡

IntelliSense 智能提示

interleaving[ˌintə(ː)ˈliːviŋ] *n.*交叉，交错

intern[inˈtəːn] *n.*实习生

internship[ˈintəːnʃip] *n.*实习（职位）

interpreter[inˈtəːpritə] *n.*解释程序

intrinsic[inˈtrinsik] *adj.*固有的，内在的

intrusion[inˈtruːʒən] *n.*闯入，侵扰

intuitively[inˈtjuitivli] *n.*直觉地，直观地

invoice[ˈinvɔis] *n.*发票，发货单

invoke[inˈvəuk] *v.*调用

irreversible[ˌiriˈvəːsəbl] *adj.*不可改变的，不可逆的

iterator[ˈitəreitə] *n.*迭代器

J

jack[dʒæk] *n.*火车头

jargon[ˈdʒɑːgən] *n.*行话

jeopardize['dʒepədaiz] v.危害

jet[dʒet] n.喷气机

job[dʒɔb] n.作业

jurisdiction[.dʒuəris'dikʃən] n.司法，权限

K

keep in mind 记住

keep up with 赶上，不落后于

keypunch['ki:pʌntʃ] n.键盘穿孔机

keystroke['ki:strəuk] n.按键

kick in 踢掉砸开

L

laptop computer 膝上型电脑

latency['leitənsi] n.延迟，等待时间

launch[lɔ:ntʃ] v.发动，开始

launder['lɔ:ndə] v.洗黑钱

law[lɔ:] n.规则，法则，规律

layer['leiə] v.把…形成或分成层次

layman['leimən] n.外行

layout['lei,aut] n.布置，安排

lean[li:n] adj.简洁的，直接的

legacy['legəsi] n.遗产，遗留物

legitimate[l i'dʒitimit] adj.合法的

leverage['li:vəridʒ] v.杠杆作用，好像通过杠杆作用进行影响

liberal['libərəl] v.自由的，大量的

lift[lift] v.解除

light up 点燃，照亮

liken['laikən] v.把…比作

line-of-sight 视线，瞄准线

lithium['liθiəm] n.锂

live[liv] adj.最新的

lock-in 锁定

loiter['lɔitə] v.闲逛，徘徊

loom[lu:m] v.迫在眉睫

loose-knit 可拆开的

lore[lɔ:] n.口头传说

low-end 低端的

loyalty['lɔiəlti] n.忠诚，忠实

M

Macintosh['mækin,tɔʃ] n.Apple公司于1984年推出的一种系列微机，简称Mac

macro['mækrəu] n.宏指令

magnitude['mægnitju:d] n.量级

mainframe['meinfreim] n.主（计算）机

malfunction[mæl'fʌŋkʃən] n.故障

malicious[mə'liʃəs] adj.恶意的

malpractice['mæl'præktis] n.玩忽职守，业务技术事故

malware 恶意软件

managed code 托管代码

manager['mænidʒə] n.管理器

mandatory['mændətəri] n.强制的，托管的

mangle['mæŋgl] v.破坏，毁损

manifest['mænifest] n.显示，清单

manipulate[mə'nipjuleit] v.操作，处理

mapping['mæpiŋ] n.映射

mark[mɑ:k] n.标志值

mashup['mæʃʌp] n.糅合

MasterCard 万事达信用卡

materialize[mə'tiəriəlaiz] v.使成真，实现

mechanics[mi'kæniks] n.力学

memory['meməri] n.内存

mentor['mentɔ:] v.指导

mesh[meʃ] n.网状物

message['mesidʒ] v.即时通讯

metadata['metə'deitə] n.元数据

metamorphic[.metə'mɔ:fik] adj.变形的，变质的

metaphor['metəfə] n.隐喻，暗喻

method['meθəd] n.办法，方法

microcontroller[maikrəukən'trəulə] n.微控制器

microprocessor[maikrəu'prəusesə] n.微处理器

middleware['midlweə] n.中间件

mind[maind] v.注意，记住

mock-up 模型，原型

model['mɔdl] v.建模

modem['məudəm] n.调制解调器

modest['mɔdist] v.适度的，适中的

modularity[.mɔdju'læriti] n.模块性

monolithic[,mɔnə'liθik] *n.*单片电路，单块集成电路
monolithic[,mɔnə'liθik] *adj.*庞大的，完整的
monopoly[mə'nɔpəli] *n.*垄断，专利权
motherboard['mʌðəbɔ:d] *n.*主板，母板
Mount Everest 珠穆朗玛峰
mouse[maus] *n.*鼠标
mug shot 大头照，面部特写
multiplex['mʌltipleks] *adj.*多路传输的
multi-stage 多级的
multitask['mʌlti,ta:sk] *v.*做多重工作
multithreading['mʌlti'θrediŋ] *n.*多线程
multitude['mʌltitju:d] *n.*大量
munitions[mju(:)'niʃəns] *n.*军需品，战争物资

N

nanotechnology['nænəutek'nɔlədʒi] *n.*纳米技术
native['neitiv] *adj.*标准的，专属的，本地的
necessitate[ni'sesiteit] *v.*使成为必需，迫使
networking card 网卡
niche[nitʃ] *n.*合适的环境，产品或服务所需的特殊
  领域
nightfall['naitfɔ:l] *n.*黄昏，傍晚
non-volatile['nɔn'vɔlətail] *adj.*非易失性的
not yet 还没有
note[nəut] *n.*注解，注释
note[nəut] *v.*着重提到，表明，指出
notebook computer 笔记本电脑
noteworthy['nəutwə:ði] *adj.*显著的，引人注意的
nothing less than 恰恰是
notification[,nəutifi'keiʃən] *n.*通知，布告
notion['nəuʃən] *n.*概念，看法
novice['nɔvis] *n.*新手，初学者
number['nʌmbə] *v.*列入，把…算作

O

object['ɔbdʒikt] *n.*对象
obviate['ɔbvieit] *v.*避免
offload[ɔf'ləud] *v.*卸载
offshore['ɔ(:)fʃɔ:] *adj.*离岸的，国外的
on the fly （计算机）运行中，匆忙地，忙禄地
on the heels of 紧跟着

on-access scanning 存取时扫描
onetime['wʌntaim] *adj.*从前的，以前的
opaque[əu'peik] *adj.*不透射的，不传导的
open source 开源
open up 开始
operating system 操作系统
opt for 选择
optimize['ɔptimaiz] *v.*使最优化
opt-out 宣布放弃选择（权）
or otherwise 或者相反
order['ɔ:də] *n.*订单
ore[ɔ:(r)] *n.*矿石
orthogonal[ɔ:'θɔgənl] *adj.*正交的
out[aut] *adj.*过时的
outage['autidʒ] *n.*运转中断，停用
outfit['autfit] *v.*配备，装备
outperform[autpə'fɔ:m] *v.*优过，胜于
outsize['aut,saiz] *adj.*特大的，极广阔的
outsource['aut,sɔ:s] *v.*外界供应，外包
overestimate['əuvə'estimeit] *v.*过高评价，过高估
  计
overlap['əuvə'læp] *v.*重叠，覆盖
override[,əuvə'raid] *v.*优先于，压倒，使无效
owe to 把…归功于

P

packet['pækit] *n.*包
packet-switched 包交换
palmtop[pa:mtɔp] *n.*掌上电脑
paper['peipə] *n.*论文，文章
paradigm['pærədaim] *n.*范例
paradoxically[,pærə'dɔksikəli] *adv.*荒谬地，自相矛
  盾地
parallel['pærəlel] *v.*匹配，与…相应
parallel ports 并行端口
parse[pa:z] *v.*解析
part[pa:t] *v.*分开，分离
participatory[pa:'tisipeitəri] *adj.*供人分享的，提供
  参加机会的
partition[pa:'tiʃən] *n.*分区，部分
partition[pa:'tiʃən] *v.*划分，把…分成部分

patch[pætʃ] v.修补，打补丁

patent['peitənt] n.专利权

payload['pei,ləud] n.有效载荷

peer-to-peer 对等，对等网络

penalty['penlti] n.损失

perceive[pə'si:v] v.察觉

perception[pə'sepʃən] n.感知，理解

peripheral[pə'rifərəl] n.外围设备

pervasive[pə:'veisiv] adj.普及的

petaflop[petəflɔp] n.每秒千万亿次浮点运算
　(FLOP: FLoating point Operations Per Second)

phase out 使逐步淘汰，逐渐停止

phish['fiʃ] v.网络钓鱼

phisher[fiʃə] n.网络钓鱼者

photonics[fəu'tɔniks] n.光子学

pick-up 获得

pico-['paikəu] adj.兆分之一

picturesque[,piktʃə'resk] adj.独特的

piggy-back 背负式装运

pin[pin] n.针，管脚，引线

pinnacle['pinəkl] n.高峰，尖端

pipe[paip] v.传送

pipeline['paip,lain] n.管道

pipelining['paip,lainiŋ] n.流水线操作

pixelization[piksəlai'zeiʃən] v.像素化

placement['pleismənt] n.放置，布局

plague[pleig] v.困扰

planar['pleinə] adj.平面的

plug-in[plʌg'in] n.插件程序

plumbing[plʌmiŋ] adj.了解的，查明的，测量的

podcast['pɔdka:sl] n.播客

polymorphism[,pɔli'mɔ:fizəm] n.多态

pop up 弹出

population[,pɔpju'leiʃən] n.种群

port[pɔ:t] v.移植

port[pɔ:t] n.端口

portable['pɔ:təbl] adj.便携式的，易携带或移动的

portable media player 便携式媒体播放器

portal['pɔ:təl] n.门户

portray[pɔ:'trei] v.描绘

pose[pəuz] v.提出，造成

pose as 假装，冒充

power['pauə] adj.专业的，有影响力的

power-up 开机

practically['præktikəli] adv.几乎，差不多

practice['præktis] n.业务，常规工作

practitioner[præk'tiʃənə] adj.从业者，实践者

precaution[pri'kɔ:ʃən] n.预防，警惕，防范

precursor[pri(:)'kə:sə] n.先驱，前任

predate['pri:'deit] v.先于，时间上先于…

predecessor['pri:disesə] n.前任，（被取代的）原有
　事物

predominantly[pri'dɔminəntli] adj.最显著的，最有
　影响的

pre-emptive[pri:'emptiv] adj.抢先的，有先买权的

premier['premjə] adj.首要的，第一的

premise['premis] n.建筑物

premium['primjəm] adj.优质的，高价的

preponderance[pri'pɔndərəns] n.优势

prescribe[pris'kraib] v.指定，规定

presence['prezns] n.势力

press[pres] n.新闻报道，报刊

pressing['presiŋ] adj.紧迫的，迫切的

prey[prei] n.猎物，牺牲品

primitive['primitiv] n.原语

privilege['privilidʒ] v.给与…特权

proactive[,prəu'æktiv] adj.主动的，先发制人的

probe[prəub] v.探查，查明

process[prə'ses] n.进程

profusely[sprəu'fju:sli] adv.丰富地

prohibitively[prə'hibitivli] adv.高得惊人地，抑制
　购买地

proliferation[prəu,lifə'reiʃən] n.激增，扩散

prominent['prɔminənt] adj.著名的，突出的

prompt[prɔmpt] n.提示

propagate['prɔpəgeit] v.繁殖，传播

propel[prə'pel] v.推动，驱使

proponent[prə'pəunənt] n.建议者，支持者

proprietary[prə'praiətəri] n.专有的，专卖的

proprietary software 专有软件

prospect['prɔspekt] n.前景，机会

protocol['prəutəkɔl] *n.*协议
prototype['prəutətaip] *n.*原型
provision[prə'viʒən] *n.*规定，条款
proximity[prɔk'simiti] *n.*接近，亲近
punch[pʌntʃ] *n.*打孔机
puppy['pʌpi] *n.*小狗，幼犬
pursue[pə'sju:] *v.*进行，继续
pursuit[pə'sju:t] *n.*追求，追寻

Q

quad-[kwɔd] *adj.*由四部分组成的，四重的
quadratically[kwə'drætikəli] *adv.*二次地，平方地
quantum['kwɔntəm] *n.*量子，量子论
quark[kwɑ:k] *n.*夸克（理论上一种比原子更小的基本粒子）
quasi-['kweisai] *adj.*类似，准，半
quaternary[kwə'tə:nəri] *adj.*四进制的
qubit['kju:bit] *n.*量子比特

R

race[reis] *v.*与…赛跑，使全速行进
rack[ræk] *n.*支架
rack mount 机架固定件
radical['rædikəl] *adj.*根本的，基础的
readership['ri:dəʃip] *n.*（报刊、书等拥有的）读者（数）
reassure[ri:ə'ʃuə] *v.*使…恢复信心，打消…的疑虑
recast['ri:'kɑ:st] *v.*重铸，改写
reciprocity[,risi'prɔsiti] *n.*互惠
recondite[ri'kɔndait] *adj.*深奥的
recur[ri'kə:] *v.*再发生，重现
recursive[ri'kə:siv] *adj.*递归的
redundant[ri'dʌndənt] *adj.*冗余的
refactor['ri:'fæktə] *v.*重构
refactoring['fæktəriŋ] *n.*重构
refer to 查阅，参考
Registry['redʒistri] *n.*注册表
regulator['regjuleitə] *n.*管理者
release[ri'li:s] *v.*发布
relegate['religeit] *v.*把…降级，把…归类，托付

reliability[ri,laiə'biliti] *n.*可靠性
relieve[ri'li:v] *v.*为…提供帮助或援助
remainder[ri'meində] *n.*剩余物
rendering['rendəriŋ] *n.*表现，渲染
repercussion[,ri:pə(:)'kʌʃən] *n.*反响
replicate['replikit] *v.*复制
repository[ri'pɔzitəri] *n.*仓库
resent[ri'zent] *v.*憎恶，怨恨
reside[ri'zaid] *v.*居住
resolution[,rezə'lju:ʃən] *n.*分辨率
retire[ri'taiə] *v.*停止使用，使报废
retriever[ri'tri:və] *n.*一种能把猎物找回来的猎犬
reveal[ri'vi:l] *v.*展现，显示
reverse-engineering 逆向工程
revoke[ri'vəuk] *v.*取消，废止
rewire[ri:'waiə] *v.*重接电线
rhetoric['retərik] *n.*修辞，花言巧语
rip[rip] *v.*用程序将（激光唱盘上的音序）存储到硬盘上
roam[rəum] *v.*漫游
robustness[rə'bʌstnis] *n.*健壮性
round-robin 一系列
router['ru:tə] *n.*路由器
royalties['rɔiəlti] *n.*版税
run[rʌn] *v.*（工作等）进行
run into 遇到，陷入
run out of 用光，用完
run up 上涨，兴起
runaway['rʌnəwei] *adj.*失控的
rush[rʌʃ] *adj.*紧急的，急迫的

S

sabotage['sæbətɑ:ʒ] *n.*蓄意破坏
scalability [,skeilə'biliti] *n.*可括缩性
scalar['skeilə] *n.*标量
scale[skeil] *v.*调节
scaled-back 按比例缩小
scam[skæm] *n.& v.*骗局，欺诈
scavenge['skævindʒ] *v.*提取有用之物
scene[si:n] *n.*舞台
scheduler['ʃedju:lə] *n.*调度程序，调节器

schema['ski:mə] *n.*模式

schematic[ski'mætik] *adj.*示意性的

scramble['skræmbl] *v.*搅乱，使混杂

scrap[skræp] *v.*扔弃

seamlessly['si:mlisli] *adv.*无缝地，连续地

secure[si'kjuə] *v.*获得

seismic['saizmik] *adj.*地震的

self-made computer组装计算机

self-tuning 自校正

sell out 脱销

semantics[si'mæntiks] *n.*语义

seminal['si:minl] *adj.*开创性的，有重大影响的

sensationalize[sen'seiʃən,laiz] *v.*加以渲染，使耸人听闻

sequencer['si:kwənsə] *n.*程序装置，定序器

serial ports 串行端口

server['sə:və] *n.*服务器

shader[ʃeidə] *n.*着色器

shipment['ʃipmənt] *n.*运送，运输

shortened form 简称，简写

show up 出现

shred['ʃred] *v.*切碎

shrink wrap 作为套装转件出套

shuffle['ʃʌfl] *v.*搬移

siege[si:dʒ] *n.*围城，围攻

sift[sift] *v.*筛选

sign up 经报名（或签约）获得

signature['signitʃə] *n.*特征代码

silo['sailəu] *n.*筒仓，地窖

simulation[,simju'leiʃən] *n.*仿真，模拟

sinister['sinistə] *adj.*险恶的

sink[siŋk] *n.*吸附物，槽

sky-rocket 像火箭一样冲天

slate[sleit] *v.*安排，指定

slot[slɔt] *n.*插槽

snoop[snu:p] *v.*窥探，偷窃

socket['sɔkit] *n.*孔，插座

solder['sɔldə] *v.*焊接

solicit[sə'lisit] *v.*恳求获得

sound[saund] *adj.*正确的，可靠的，合理的

source code 源代码

spam[spæm] *n.*垃圾邮件

spammer[spæmə] *n.*发垃圾邮件的人

span[spæn] *v.*跨越

spatial['speiʃəl] *adj.*空间的

specification[,spesifi'keiʃən] *n.*规格，规约，说明书

spectrum['spektrəm] *n.*光谱

speculative['spekjulətiv] *adj.*推测的

speedup['spi:dʌp] *n.*加速

spin[spin] *v.*使快速旋转

sponsor['spɔnsə] *v.*资助，赞助

spoof[spu:f] *v.*哄骗，戏弄

sport[spɔ:t] *v.*浪费

spot[spɔt] *v.*发现，准确地定出…的位置

sprout[spraut] *v.*萌芽，迅速成长和出现

spur[spə:] *v.*刺激，激励，鞭策

spurious['spjuəriəs] *adj.*伪造的，假造的

spyware 间谍软件

stack[stæk] *n.*堆栈

staging area 临时数据交换区

stall[stɔ:l] *v.*停止，迟延

stamp[stæmp] *n.*印章，戳记

staple['steipl] *n.*主要产品，日常必需品

start off 出发，动身

start with 首先，第一

state of the art 最新水平

steer[stiə] *v.*驾驶

stem[stem] *v.*抵抗，阻止

step in 介入，走进

stock[stɔk] *n.*坐（式）

storage['stɔridʒ] *n.*存储，存储器

stored procedure 存储过程

storefront[stɔ:frʌnt] *n.*店头，店面

streamline['stri:mlain] *v.*使简单化

strip-down 脱去，拆开

stuff[stʌf] *v.*塞满，填充

stumble['stʌmbl] *v.*跌绊

sublime[sə'blaim] *n.*顶点

subroutine[,sʌbru:'ti:n] *n.*子程序

subscriber[sʌbs'kraibə] *n.*订购者，用户

substantially[səb'stænʃ(ə)li] *adv.*基本上

subversion[sʌb'və:ʃən] *n.*颠覆，破坏

successor[sək'sesə] *n.*后继者，后续的事物

suite[swi:t] *n.*套装，套件

supercomputer[,sju:pəkəm'pju:tə] *n.*超级计算机

superconductor[ˌsjuːpəkənˈdʌktə] *n.*超导体
superimposition[ˈsjuːpərˌimpəˈziʃən] *n.*叠印
superposition[ˌsjuːpəpəˈziʃən] *n.*重叠
superscalar[ˈsjuːpəˈskeilə] *n.*超标量体系结构
surf[səːf] *v.*冲浪
surge[səːdʒ] *n.*电涌
surveillance[səːˈveiləns] *n.*监视
susceptibility[səˌseptəˈbiliti] *n.*易感染性
sustained[səsˈteind] *adj.*持续不变的，相同的
switch[switʃ] *n.*交换机
synchronize[ˈsiŋkrənaiz] *v.*同步
syndication[ˈsindikeiʃən] *n.*整合，聚合
synonym[ˈsinənim] *n.*同义词
synonymous[siˈnɔniməs] *adj.*同义的
syntax[ˈsintæks] *n.*语法

T

tabular[ˈtæbjulə] *adj.*表格式的
tailor[ˈteilə] *v.*裁剪
take into account 考虑
take up 占空间
tamper[ˈtæmpə] *v.*篡改
target[ˈtɑːgit] *v.*把…作为目标
tarnish[ˈtɑːniʃ] *v.*减损，使…成为泡影
task[tɑːsk] *v.*分派任务
tearing[ˈtɛəriŋ] *adj.*令人难受的，猛烈的
terabyte[ˈterəbait] *n.*1000GB，万亿字节
terminal[ˈtəːminl] *n.*终端
textual[ˈtekstjuəl] *adj.*原文的，正文的，逐字的
texture[ˈtekstʃə] *n.*纹理
theorem[ˈθiərəm] *n.*定理，法则
third-party 第三方
thorny[ˈθɔːni] *adj.*棘手的，麻烦的
thread[θred] *n.*线程
three-dimensional 立体的，三维的
throughput[ˈθruːput] *n.*吞吐量
thwart[θwɔːt] *v.*阻止…的发生
tidal[ˈtaidl] *adj.*潮水般的
tide[taid] *n.*高潮
tight-lipped 紧闭嘴巴的，沉默的
tinker[ˈtiŋkə] *v.*笨手笨脚地做事，修补

tip[tip] *n.*指点，提示
to date 到目前为止
token ring 令牌环网
toolkit[tuːlkit] *n.*工具包，工具箱
tooltips 工具提示
topology[təˈpɔlədʒi] *n.*拓扑
tout[taut] *v.*吹捧
tradeoff 折衷，权衡
transaction[trænˈzækʃən] *n.*事务，交易
transistor[trænˈzistə] *n.*晶体管
transparent[trænsˈpɛərənt] *adj.*透明的
trick[trik] *v.*哄骗，欺诈
trigger[ˈtrigə] *n.*触发
trillion[ˈtriljən] num.万亿
trojan horse 特洛伊木马
troubleshoot[ˈtrʌblʃuːt] *v.*检修
trump[trʌmp] *v.*胜过
twisted pair 双绞线
type-safe 类型安全
typify[ˈtipifai] *v.*代表

U

ubiquity[juːˈbikwəti] *adv.*到处存在，普遍存在
unauthentic[ˈʌnɔːˈθentik] *adj.*不可靠的，不可信的
uncrackable[ʌnˈkrækəbl] *adj.*
underdog[ˈʌndədɔg] *n.*失败者，受压迫者
unmanaged code 非托管代码，未受管理（或控制）的代码
unpack[ˈʌnˈpæk] *v.*打开取出
unprecedented[ʌnˈpresidəntid] *adj.*空前的，史无前例的
unravel[ʌnˈrævəl] *v.*弄清阐明，解决
unruly[ʌnˈruːli] *adj.*难以控制的
unveil[ʌnˈveil] *v.*公布
up and down 到处，详细
up front 在前面，预先，先期
up to date 最新的，直到最近的
upfront[ʌpfrʌnt] *adj.*在前面的，在最前面的
uptake[ˈʌpteik] *n.*理解，举起
use case 用例
utility[juːˈtiliti] *n.*实用程序，应用程序

utopian[juːˈtəupjən] *adj.*乌托邦的，理想化的

V

vacuum tube 真空管，电子管
value[ˈvæljuː] *v.*评价
vane[vein] *n.*风向标
vector[ˈvektə] *n.*力量，动力
vein[vein] *n.*矿脉，特色，风格
vested[ˈvestid] *adj.*既定的，确定的
viable[ˈvaiəbl] *adj.*切实可行的，可实施的
vice versa 反之亦然
view[vjuː] *n.*视图
virus[ˈvaiərəs] *n.*病毒
vulnerability[ˌvʌlnərəˈbiləti] *n.*弱点

W

wane[wein] *v.*衰落
Web site 网站

widget[ˈwidʒit] *n.* 窗口小部件，小工具
wiki[wiki] *n.*维基
wipe out 消灭
with regard to 关于，对于
with the purpose of 以…为目的
withhold[wiðˈhəuld] *v.*拒给，保留
word-of-mouth 口碑
workflow[ˈwəːkfləu] *n.*工作流
workstation[ˈwəːksteiʃən] *n.*工作站
worm[wəːm] *n.*计算机网络"蠕虫"
write off 一口气写成

Y

yield[jiːld] *n.*产量，收益

Z

zealous[ˈzeləs] *adj.*热心的，积极的
zoom[zuːm] *v.*移向（或移离）目标

Abbreviations

A

AC	Alternating Current	交流电
ACID	Atomicity, Consistency, Isolation and Durability	数据库事务正确执行的四个基本要素：原子性、一致性、隔离性、持久性
Ajax	Asyndironous JavScript and XML	异步JavaScript和XML
AKA	Also Known As	又名，也称
ALM	Application Life-cycle Management	应用程序生命周期管理
ANSI	American National Standards Institute	美国国家标准协会
API	Application Programming Interface	应用编程接口
ARCNET	Attached Resource Computer NETwork	附加资源计算机网络
ARPA	Advanced Research Projects Agency	（美国国防部）高级研究计划署
ARPANET	Advanced Research Projects Agency Network	（美国）高级研究计划署网络
ATM	Automated Teller Machines	自动柜员机

B

B2B	Business-to-Business	企业对企业的电子商务模式
B2C	Business-to-Customer	企业对消费者的电子商务模式
B2G	Business-to-Government	企业对政府的电子商务模式
BBS	Bulletin Board System	电子公告栏系统
BCL	Base Class Library	基类库
BIOS	Basic Input Output System	基本输入输出系统
BLOB	Binary Large OBject	二进制大对象
BNC	Bayonet Neill-Concelman	尼尔–康塞曼插刀（一种插口）
BOINC	Berkeley Open Infrastructure for Network Computing	伯克利开放式网络计算平台
BSD	Berkeley Software Distribution	伯克利软件套件

C

CAD	Computer-Aided Design	计算机辅助设计
CD	Compact Disk	光盘
CD-R	Compact Disc-Recordable	可记录光盘
CD-ROM	Compact Disc Read-Only Memory	光盘驱动器
CD-RW	Compact Disc ReWritable	可擦写光盘

CGA	Color Graphics Array	彩色图形阵列
CiD	Commission for International Development	国际开发委员会
CIL	Common Intermediate Language	通用中间语言
CLI	Common Language Infrastructure	通用语言基础结构
CLR	Common Language Runtime	公用语言运行时
CMM	Capability Maturity Model	能力成熟度模型
CMMI	Capability Maturity Model Integration	能力成熟度模型集成
CMP	Chip-level MultiProcessor	芯片级多处理器
COM	Component Object Model	组件对象模型
CPU	Central Processing Unit	中央处理器
CSMA/CD	Carrier Sense Multiple Access/Collision Detection	载波侦听多路访问/冲突检测
CSS	Cascading Style Sheets	层叠样式表
CTP	Community Technology Previews	社区技术预览版
CTS	Common Type Specification	通用类型说明

D

DBA	DataBase Administrator	数据库管理员
DBMS	DataBase Management System	数据库管理系统
DC	Direct Current	直流电
DEC	Digital Equipment Corporation	美国数字设备公司
DEM	Digital Elevation Model	数字高程模型
DHCP	Dynamic Host Configuration Protocol	动态主机分配协议
DHTML	Dynamic HTML	动态HTML
DIMM	Dual In-line Memory Module	双列直插式内存模块
DIX	Digital/Intel/Xerox	数字设备/英特尔/施乐公司
DIY	Do It by Yourself	自己动手
DLL	Dynamic Link Library	动态链接库
DNA	DeoxyriboNucleic Acid	脱氧核糖核酸
DRAM	Dynamic Random Access Memory	动态随机访问存储器
DSL	Digital Subscriber Line	数字用户线路
DSP	Digital Signal Processors	数字信号处理器
DVD	Digital Video Disc	数字化视频光盘
DVD-ROM	Digital Video Disc Read-Only Memory	数字化视频光盘驱动器
DVFS	Dynamic Voltage and Frequency Scaling	动态电压与频率调节

E

EDI	Electronic Data Interchange	电子数据交换
EDPM	Electronic Data Processing Machine	电子数据处理机
EDVAC	Electronic Discrete Variable Automatic Computer	离散变量自动电子计算机
EEPROM	Electrically Erasable Programmable Read-Only Memory	电可擦除可编程只读存储器
EFT	Electronic Funds Transfer	电子资金转账

EGA	Enhanced Graphics Array	增强型图形阵列
EIDE	Enhanced IDE	增强型IDE接口
ENIAC	Electronic Numerical Integrator And Calculator[Computer]	电子数字积分计算机
EPIC	Explicitly Parallel Instruction Computing	显式并行指令计算
ERD	Entity Relationship Diagrams	实体关系图
ERMA	Electronic Recording Method of Accounting	电子账目记录方法
ERP	Enterprise Resource Planning	企业资源规划
ETL	Extraction-Transformation-Loading	数据抽取、转换和加载
Ext3	Third extended file system	一种日志式文件系统

F

FAQ	Frequently Asked Question	常见问题解答
FCC	Federal Communications Commission	美国联邦通讯管理委员会
FDDI	Fiber Distributed Data Interface	光纤分布式数据接口
FOL	First-Order Logic	一阶逻辑
FOSE	Future Of Software Engineering conference	软件工程前景讨论会
FSB	Front Side Bus	前端总线
FSM	Finite State Machines	有限状态机
FTP	File Transfer Protocol	文件传输协议

G

GIS	Geographic Information System	地理信息系统
GNU	GNU's Not Unix	一个完全由自由软件组成的计算机操作系统
GNU GPL	GNU General Public License	GNU通用公共许可证
GPU	Graphics Processing Unit	图形处理单元
GUI	Graphical User Interface	图形用户界面

H

| HTML | HyperText Markup Language | 超文本标记语言 |

I

I/O	Input/Output	输入/输出
IA	Intel Architecture	英特尔体系结构
IBM	International Business Machines	（美国）国际商用机器公司
ICSE	International Conference on Software Engineer	软件工程国际会议
IDE	Intelligent Drive Electronics or Integrated Drive Electronics	集成电路设备，智能磁盘设备
IDE	Integrated Drive Electronics	电子集成驱动器
IEEE	Institute of Electrical and Electronics Engineers	美国电气和电子工程师协会

IFF	Identification Friend or Foe	敌我识别
IL	Intermediate Language	中间语言
IMD	International Institute for Management Development	国际管理发展研究院
IMP	Interface Message Processor	接口报文处理器
INWG	International Network Working Group	国际互联网络工作组
IPC	Instruction Per Cycle	每周期指令
IPO	Initial Public Offering	首次公开发行股票
ISO	International Organization for Standardization	国际标准化组织
ISO	ISOlation	一种镜像文件
ISV	Independent Software Vendors	独立软件开发商

J

JIT	Just In Time	实时，激活
JRE	Java Runtime Environment	Java运行环境
JSON	JavaScript Object Notation	JavaScript标识对象的方法
JVM	Java Virtual Machine	Java虚拟机

K

KML	Keyhole Markup Language	Keyhole标记语言，是一个基于XML语法和文件格式的文件，用来描述和保存地理信息如点、线、图片、折线并在Google Earth客户端之中显示

L

LAMP	Linux, Apache, MySQL and PHP	一组常用来搭建动态网站或者服务器的开源软件
LAN	Local Area Network	局域网
LGPL	Lesser General Public License	较宽松通用公共许可证
LINQ	Language INtegrated Query	语言级集成查询
LMI	Linux Mark Institute	Linux商标协会

M

MAC	Media Access Control	介质访问控制
malware	Malicious Software	恶意软件
MFC	Microsoft Foundation Classes	微软基础类
MICR	Magnetic Ink Character Recognition	磁墨水字符识别
MIT	Massachusettes Institute of Technology	（美国）麻省理工学院
MPEG	Moving Pictures Experts Group	动态图像专家组，视频、音频、数据的压缩标准

MPU	MicroProcessor Unit	微处理器
MRO	Maintenance, Repair and Operations	维护、维修、运行
MSDN	Microsoft Developer Network	微软开发者网络
MS-DOS	Microsoft Disk Operating System	微软磁盘操作系统
Muni-Fi	Metropolitan-wide Wi-Fi	城域Wi-Fi网

N

NASA	National Aeronautics and Space Administration	（美国）国家航空航天局
NATO	North Atlantic Treaty Organization	北大西洋公约组织
NIC	Network Interface Card	网络接口卡
NTFS	New Technology File System	Windows NT以上版本支持的一种文件系统
NVRAM	Non-Volatile RAM	非易失性随机访问存储器

O

ODBC	Open DataBase Connectivity	开放数据库互连
OEM	Original Equipment Manufacturer	原始设备制造商
OFDM	Orthogonal Frequency Division Multiplexing	正交频分复用技术
OLAP	On-Line Analysis Processing	联机分析处理
OMG	Object Management Group	对象管理组织
OMT	Object Modeling Technique	对象建模技术
OOP	Object-Oriented Programming	面向对象的程序设计
OS	Operating System	操作系统
OWL	Web Ontology Language	网络本体语言

P

PARC	Palo Alto Research Center	（施乐公司）帕洛阿尔托研究中心
PC	Personal Computer	个人计算机
PCB	Printed Circuit Board	印制电路板
PDA	Personal Digital Assistant	个人数字助理
PNG	Portable Network Graphic	可移植的网络图像文件格式
POSIX	Portable Operating System Interface of Unix	Unix可移植操作系统接口
POST	Power On Self Test	通电自检测试

R

RAID	Redundant Array of Independent Disk	独立冗余磁盘阵列
RAISE	Rigorous Approach to Industrial Software Engineering	工业软件工程的严格方法
RAM	Random-Access Memory	随机访问存储器
RDBMS	Relational DataBase Management System	关系型数据库管理系统
REST	REpresentational State Transfer	表象化状态转换

RF	Radio Frequency	无线射频
RHEL	Red Hat Enterprise Linux	Red Hat 企业版 Linux
ROM	Read-Only Memory	只读存储器
RSS	Really Simple Syndication	聚合新闻服务
RUP	Rational Unified Process	Rational 统一过程

S

SATA	Serial Advanced Technology Attachment	串行高技术附件
SCSI	Small Computer Systems Interface	小型计算机系统接口
SDN	Sun Developer Network	Sun开发者网络
SE	Software Engineering	软件工程
SIMD	Single Instruction Multiple Data	单指令多数据流
SLA	Service-Level Agreement	服务等级协议
SMP	Symmetric MultiProcessing	对称多处理
SoC	System-on-a-Chip	单片系统
SP	Service Pack	服务包，补丁
SQL	Structured Query Language	结构化查询语言
SQLOS	SQL Operating System	SQL操作系统
SRTM	Shuttle Radar Topography Mission	航天飞机雷达地形测绘任务
SSIS	SQL Server Integration Services	SQL Server集成服务
SSL	Security Socket Layer	安全套接层
STL	Standard Template Library	标准模板库
SUS	Single UNIX Specification	单一UNIX规范

T

TBB	Threading Building Block	线程构建模块
TCP	Transfer Control Protocol	传输控制协议
TCP/IP	Transmission Control Protocol/Internet Protocol	传输控制协议/因特网协议
TDS	Tabular Data Stream	表格数据流
T-SQL	Transact-SQL	事务型SQL

U

UAC	User Account Control	用户账户控制
UML	Unified Modeling Language	统一建模语言
USB	Universal Serial Bus	通用串行总线
Usenet	Uses Network	新闻讨论组
UTP	Unshielded Twisted Pair	非屏蔽双绞线

V

VBS	Visual Basic Script	VB脚本
VDM	Vienna Development Method	维也纳开发方法

VLIW	Very-Large Instruction Word	超长指令字
VM	Virtual Machine	虚拟机
VMS	Virtual Memory System	虚拟内存系统
VO	Virtual Organization	虚拟组织
VoWLAN	Voice over WLAN(Wireless Local Area Networks)	基于无线局域网络的IP语音通信
VPN	Virtual Private Network	虚拟专用网络

W

WAN	Wide Area Network	广域网
WEP	Wired Equivalent Privacy	有线对等保密
WLAN	Wireless Local Area Network	无线局域网络
WMM	Wireless MultiMedia	无线多媒体
WMS	Web Map Service	网络地图服务
WPA	Wi-Fi Protected Access	Wi-Fi保护访问
WPF	Windows Presentation Foundation	微软用于Windows的统一显示子系统
WVOIP	Wireless VoIP(Voice over IP)	无线VoIP

X

XAML	eXtensible Application Markup Language	可扩展应用程序标记语言
XHTML	eXtensible HyperText Markup Language	可扩展超文本标记语言
XML	eXtensible Markup Language	可扩展标记语言
XP	Extreme Programming	极限编程
XSL	eXtensible Stylesheet Language	可扩展样式表语言
XSLT	XSL Transformations	XSL 转换